SOCIAL WORK IN PUBLIC HEALTH AND HOSPITALS

SOCIAL WORK IN PUBLIC HEALTH AND HOSPITALS

Sharon Duca Palmer, CSW, LMSW

School Social Worker, ACLD Kramer Learning Center,
Bay Shore, New York; Certified Field Instructor,
Adelphi University School of Social Work,
Garden City, New York, U.S.A.

Apple Academic Press

Social Work in Public Health and Hospitals

First Published in the Canada, 2011
Apple Academic Press Inc.
3333 Mistwell Crescent
Oakville, ON L6L 0A2
Tel. : (888) 241-2035
Fax: (866) 222-9549
E-mail: info@appleacademicpress.com
www.appleacademicpress.com

The full-color tables, figures, diagrams, and images in this book may be viewed at www.appleacademicpress.com

First issued in paperback 2021

ISBN 13: 978-1-77463-249-9 (pbk)
ISBN 13: 978-1-926692-85-2 (hbk)

Sharon Duca Palmer, CSW, LMSW

Cover Design: Psqua

Library and Archives Canada Cataloguing in Publication Data
CIP Data on file with the Library and Archives Canada

CONTENTS

INTRODUCTION

Social work is a difficult field to operationally define, as it is practiced differently in many settings. It is a very diverse occupation and one that can be practiced in settings such as hospitals, clinics, welfare agencies, schools, and private practices.

The main goal of all social work practice is to assist the client to function at the best of their ability and assess what their needs are. Social workers help clients with problem-solving strategies, such as defining personal goals, focusing on what is necessary to make changes, and helping them through the process.

Social work is a demanding field and is often emotional draining. Many social workers have large caseloads, limited resources for their clients, and often work for relatively low salaries. But the personal rewards can be very satisfying.

The social work profession is committed to promoting social and economic policy though helping to improve people's lives. Research is conducted to improve social services, community development, program evaluation, and public administration. The importance of research in these areas is to examine variables that can be addressed in order to resolve issues. Research can lead to what is called "best practice". By utilizing "best practice", a social worker is engaging clients based on research that is intended to increase successful outcomes.

Social work is one of the most diverse careers available. Most social workers are employed by health care facilities and government agencies. These facilities can include hospitals, mental health clinics, nursing homes, rehabilitation centers, schools, child welfare agencies, and private practice.

Social work's interface with mental health promotion and the treatment of mental illness dates to the earliest roots of our profession. While many social workers provide mental health services in private practice settings, the majority of services are offered in community-based agencies, both public and private, and in hospitals and prisons. Social workers are the largest provider of mental health services, providing more services than all other mental health care providers combined. These workers also often provide services to those who are struggling with substance abuse.

Twenty-first century health issues are complex and multidimensional, requiring innovative responses across professions at all levels of society. Public health social workers work to promote health in hospitals, schools, government agencies and local community-based settings, making connections between prevention and intervention from the individual to the whole population.

In an ideal world, every family would be stable and supportive. Every child would be happy at home and at school. Every elderly person would have a carefree retirement. Yet in reality, many children and families face daunting challenges. For example, single parents struggle to raise kids while working. Teens may become parents before they are ready. Child social workers help kids get back on track so they can lead healthy, happy lives.

Rapid aging populations are expected worldwide. With the rapid growth of this population, social work education and training specializing in older adults and practitioners interested in working with older adults are increasingly in demand. Geriatric social workers typically provide counseling, direct services, care coordination, community planning, and advocacy in an array of organizations including in homes, neighborhoods, hospitals, senior congregate living and nursing facilities. They work with older people, their families and communities, as well as with aging-related policy, and aging research

In whatever subcategory they work, social workers help provide support services to individuals and communities by assessing their needs in order to improve the quality of life and overall well-being. This can lead to positive changes in people's environments, dignity, and self-worth. It can also lead to changes in social policy for those who are vulnerable and oppressed. Social workers change entire communities for the better.

There have been many changes emerging in the social work profession. The uses of the Internet and online counseling have been major trends. Some people are more likely to seek assistance and information first through the use of the Internet. There has also been a strong move for collaborating between professions when providing services in order to offer clients more options for success. Keeping up to date with best practice research, licensing requirements, continuing education, and professional ethics make this an exciting and challenging time to be a social worker!

— Sharon Duca Palmer, CSW, LMSW

Direct Costs Associated with the Appropriateness of Hospital Stay in Elderly Population

Joaquín F. Mould-Quevedo, Carmen García-Peña,
Iris Contreras-Hernández, Teresa Juárez-Cedillo,
Claudia Espinel-Bermúdez, Gabriela Morales-Cisneros
and Sergio Sánchez-García

ABSTRACT

Background

Ageing of Mexican population implies greater demand of hospital services. Nevertheless, the available resources are used inadequately. In this study, the direct medical costs associated with the appropriateness of elderly populations hospital stay are estimated.

Methods

Appropriateness of hospital stay was evaluated with the Appropriateness Evaluation Protocol (AEP). Direct medical costs associated with hospital stay under the third-party payer's institutional perspective were estimated, using as information source the clinical files of 60 years of age and older patients, hospitalized during year 2004 in a Regional Hospital from the Mexican Social Security Institute (IMSS), in Mexico City.

Results

The sample consisted of 724 clinical files, with a mean of 5.3 days (95% CI = 4.9–5.8) of hospital stay, of which 12.4% (n = 90) were classified with at least one inappropriate patient day, with a mean of 2.2 days (95% CI = 1.6 – 2.7). The main cause of inappropriateness days was the inexistence of a diagnostic and/or treatment plan, 98.9% (n = 89). The mean cost for an appropriate hospitalization per patient resulted in US$1,497.2 (95% CI = US$323.2 – US$4,931.4), while the corresponding mean cost for an inappropriate hospitalization per patient resulted in US$2,323.3 (95% CI = US$471.7 – US$6,198.3), (p < 0.001).

Conclusion

Elderly patients who were inappropriately hospitalized had a higher rate of inappropriate patient days. The average of inappropriate patient days cost is considerably higher than appropriate days. In this study, inappropriate hospital-stay causes could be attributable to physicians and current organizational management.

Background

Ageing of the Mexican population is one of the higher impact phenomena which began manifesting during the XX Century, and which will undoubtedly be an essential element in the creation of the history of Mexico in the XXI Century. Life expectancy of Mexicans doubled during the second half of the last century, raising from 36 years in 1950 to 74 years in year 2000, and it is expected to reach 80 years in 2050 [1].

Ageing of the population will imply a greater demand of health services, as a consequence of the high rates associated morbidity. Among the diseases of the elderly population are chronic-degenerative conditions, that make them fragile and turns them into a population that utilize many health services. However, it is well known that part of the hospital resources are used inadequately, either because the patients receive assistance that does not turn into health benefits or because care services could be provided at different institutional levels representing lower costs [2,3].

Evaluation of resource use in hospital health systems allows to establish the necessary actions which will correct the identified organizational problems. Unjustified hospital admissions and stays of elderly patients, do not only increase costs but are also related to poor health service and higher mortality resulting from several complications that come together, for instance, hospital infections, pressure ulcers or venous thrombosis, among others. Thus, the permanent evaluation of hospital service utilization is an essential topic that must be considered to improve resource assignment and to increase the quality of medical assistance in institutions that render services [4,5].

There are several reports that have calculated inappropriate utilization figures. However, very few have studied the associated factors [6-8].

Finally, the economic analysis of health services provided to the elderly population was done in general terms by other researchers [9,10], concluding that the associated costs including medical assistance required by the elderly are higher in comparison to the rest of the population. Then again, the cost of appropriate and inappropriate hospital stay in the elderly population has not been identified, at least in developing countries where medical practice vary and could have important effects on the hospital budgets in comparison with other developed countries. Even if medical costs would be expected to be lower in developing countries this expenses could have a significant impact over health budgets. Previous studies of inappropriate hospital length of stay showed that this phenomenon is mainly generated by doctors whom are not required to justify individual hospital stay, and hence there is an incentive to prolong hospitalization when there are empty beds [11,12]. In addition, other researchers identified a number of key causes of inappropriate admissions and lengths of stay, including: the limited capacity of health and social care resources; poor communication between primary and secondary care clinicians and the cautiousness of clinicians who manage patients in community settings [13]. In this sense, economic literature have been elaborated in order to estimate direct medical costs of length of stay and to estimate budget impact of inappropriate hospitalizations [14,15]. The aim of this study was to estimate direct medical costs associated with inappropriateness hospitalizations days in the elderly population within a representative Mexican hospital.

Methods

A retrospective study was conducted, reviewing clinical files of 60 years of age and older patients (n = 7,540) admitted from January 1st to December 31st of 2004 within the Mexican Social Security Institute (IMSS).

The IMSS was created by law in 1943 and is funded by the government, employers, and employees. It is a social security system; therefore, the only requirement to be registered is to be employed, regardless of one's state of health. Workers, their parents, and other close relatives are assigned to a Family Medicine Unit, which is the primary health care provider. The IMSS offers a comprehensive package of benefits that include health care services at all levels of care and economic benefits such as a pension. Medical services are provided by levels of care and a reference system. Mexico City has a population of nearly 860,000 adults aged 60 years and older, 418,000 of whom (48.6%) are affiliated to the IMSS. IMSS has 16 general hospitals in Mexico City. From those, the General Hospital "Dr. Carlos Mac Gregor Sánchez Navarro" was selected. This hospital is one of the biggest hospitals in Mexico City and it has 278 total beds. A total of 465 physicians and 705 nurses work there. General services as internal medicine, surgery, intensive care, pediatrics, plus some specialized services such as hematology, nephrology integrate the services provided.

The size of the sample was estimated assuming that 10% of the hospital admissions are inappropriate [16], with an accuracy of 0.02 and a reliability level of 95%. The required sample was of 863 files, plus a 20% considering the exclusion criteria. The files were selected simply by random.

We excluded files of patients with voluntary hospital discharges, patients with illegible files, lacking information necessary for evaluation, files of patients transferred to other hospitals and those that were not found when the information was collected.

The unit of analysis was made up of the admission process clinical hospital file registries, first days and subsequent patient days to admission, with the exception of the medical discharge day. The version adapted to Spanish and validated in the elderly population by our Appropriateness Evaluation Protocol (AEP) group was used as a tool to determine hospitalization appropriateness of the study subjects [17-19]. The AEP was previously validated in elderly Mexican population in 2005 [16]. The first AEP set comprised 16 criteria that establish the need for hospitalization on the hospital admission day. The first ten criteria were related with the severity of the patient's clinical condition, while the remaining six criteria were associated with frequency with which the provided health services were utilized. The presence of at least one of these criteria on the first admission day was enough to consider it appropriate; on the other hand, it was considered inappropriate when it did not comply with any of the criteria. In addition, to determine the need for further days of stay subsequent to the admission day -with the exception of the admission day itself-, we used a second AEP set comprising 27 criteria related with medical service rendered, nursing services and the patient's clinical condition. As in the previous case, compliance with one sole criterion was sufficient to consider the patient days reviewed as appropriate and inappropriate when it did not comply with any of the criteria. Likewise, admission causes and inappropriate patient days were reported.

A group of three nurses with bachelor degrees was previously trained in AEP handling and application. The intra-reviewer agreement of hospital admissions presented a Kappa coefficient of >80%, while patient days yielded Kappa>86%. Regarding to inter-reviewer agreement, the result was Kappa>93%, and >85% for admission and patient days, respectively.

Age and gender were also collected as well as admission service and comorbidity. Comorbidity was defined as the presence of one or more disorders (or diseases) in addition to a primary disease or disorder.

The economic analysis consisted of an estimate of direct medical costs associated with appropriate and inappropriate assistance provided to the subjects included in the study during their hospital stay. Direct medical costs include all related costs generated during the hospitalization of patients (drugs, laboratory and radiologic exams, inter-visits to other specialists, procedures, emergency and administrative expenses). Neither out-of-pocket expenses nor indirect costs (productivity losses) are included since the perspective is from the third-party payer's.

The economic analysis procedure was realized by resource identification, measurements and valuation. Identification and measurement of the resources was carried out through a review of the clinical files to generate a listing of goods and services used in every hospital stay, identifying the headings corresponding to patient days, surgeries, special procedures, laboratory tests and office examinations, medication and inter-visits to other specialists.

Cost estimates were obtained applying the unitary costs of goods and services used in every hospital stay, which were obtained from IMSS institutional data base, accessing http://www.imss.gob.mx web page, as well as from the notices of IMSS Planning and Finance Division, 2006. The costs are expressed in US Dollars according to the 2006 exchange rate officially reported by Banco de México. No discount rate was applied, since the analysis horizon was within one year.

Statistical Analysis

Of the total of identified hospital stays, values of the mean of the interest socio-demographic and clinical variables were obtained, and the cases corresponding to sex, age range, service to which the patient was admitted, appropriate admission, length of stay (in days) and number of comorbidities conditions were determined. Similarly, the reasons why the admission was considered as inappropriate and patient days were identified and estimated. The ratio of inappropriate days refers to the number of inappropriate length of stay per patient measured in days divided by the number of total patient days (appropriate plus inappropriate) by 100.

Mean inappropriate days only considering patients who presented inappropriate days.

Regarding the patient days: direct medical costs per patient with their reliability intervals were estimated at 95% (95% CI). Mean costs of appropriate and inappropriate patient days were calculated from the mean cost for each group (patients). In addition, the mean costs were compared through the t-student test for independent samples, with a 95% significance level.

Finally, a multivariate analysis was performed to find out what impact do socio-demographic factors (sex and age) and clinical factors (appropriate admission, length of stay, number of comorbidities conditions and inappropriate patient days) have on the hospital stay's total cost.

The patient days cost was transformed to a natural logarithm in the multivariate analysis to reduce the variability shown by costs estimated for patients with different pathologies and several severity levels: The latter result in heterogeneous patient days and different quantities of hospital resource use. Sex and admission variables were handled as dichotomies, while age, length of stay, number of comorbidity conditions and inappropriate days variables were handled as a continuum. Variables used in the multivariate analysis were chosen from the available factors founded in the hospital clinical files for each patient.

The Ordinary Least Squares assumption evaluation was done paying special attention to the homoscedasticity (assessed with the Breusch-Pagan and White tests), as the transversal cohort of patients included in the sample presented atypical factors, such as hospital stays that lasted more than 90 days and variation in the use of resources resulting from the diversity of reasons for medical assistance.

Ethics

The research protocol for this study (2004-3607-0009) was reviewed and approved by the Local Commission of Health Research and the Sub-Committee of Ethics of the Mexican Social Security Institute, Delegation 3 and 4 in Mexico City.

Results

General Characteristics

A total of 1,036 hospital clinical files were selected, of which 31.2% (n = 312) were excluded given that they did not fulfill the age or the admission period criteria. The final sample comprised 724 files: 51.9% females with an average age of 76.9 (± 9.2) years and the remaining 48.1% males with an average age of 73.9 (± 8.0) years.

Inappropriate Admission and Hospital Stays

The patient days mean was 5.3 (95% CI = 4.9–5.8) days, with 5.6 (95% CI = 4.9–6.3) and 5.0 (95% CI = 4.6–5.5) days for females and males, respectively.

Of the 724 hospital admissions, only 1.5% (n = 11) were classified as inappropriate. Those admitted inappropriately had more inappropriate patient days (17.1%), in comparison to those whose admissions were appropriate (5.0%). Among the 1.5% patients admitted inappropriately, who only required nurse assistance were 63.6% (n = 7). On the other hand, 27.3% (n = 3) of the patients required assistance from a hospital specialized in chronic diseases; and premature admission of one day or more previous to appointments for tests resulted in 9.1% (n = 1).

From the 724 files reviewed, 12.4% (n = 90) were classified with at least one day of inappropriate hospital stay. The mean of inappropriate patient days per patient of the 90 files was 2.2 days (95% CI = 1.6 – 2.7). The reasons for considering them as inappropriate stays were that there was no diagnostic plan and/or treatment, 98.9% (n = 89); that there was a planned discharge without written orders, 3.3% (n = 3); there was no work at the hospital those days (certain diagnostic procedures are not performed during the weekend or in holidays), 2.2% (n = 2); patient programmed for diagnostic tests or treatment (including surgery) whose appointment was cancelled due to any other reason (for instance, an emergency case is put before an elective case or essential personnel from the hospital was ill, etc.) 1.1% (n = 1); and, others 2.2% (n = 2). It is worth pointing out that five files fulfilled more than one of these causes.

Five percent (5.1%, n = 198) of the 3891 days of hospital stays were classified as inappropriate. Table 1 shows the ratio of inappropriate patient days according to the characteristic of the sample.

Direct Medical Costs and Associated Factors

Mean costs of inappropriate hospital days was calculated from the mean cost of 90 patients (files) founded in our sample. The mean cost for an appropriate hospitalization per patient resulted in US$1,497.2 (95% CI = US$323.2 – US$4,931.4), while the corresponding mean cost for an inappropriate hospitalization per patient resulted in US$2,323.3 (95% CI = US$471.7 – US$6,198.3), showing statistically significance difference among them (p < 0.001). Differences are mainly explain due to the higher number of unnecessary days the patient is treated in the hospital using healthcare resources such as additional laboratory and gabinet exams (7%), inter-visits to other specialists (28%), drugs (5%) and administrative expenses (60%). Nevertheless, an inappropriate day costs 18% less than an appropriate day due is less intensive in resource use (patients are mainly in observation not treated intensively).

Table 1. Appropriate of admission and patient days.

	Patients n (%)	Patient days n (%)	Rate of Inappropriateness days Rate (95% CI)
Total	724 (100)	3891 (100)	5.1 (3.4 – 5.9)
Sex			
Female	376 (51.9)	2118 (54.4)	5.0 (3.2 – 5.5)
Male	348 (48.1)	1773 (45.6)	5.2 (3.5 – 6.1)
Age (years)			
60-64	87 (12.0)	446 (11.5)	5.2 (3.3 – 5.8)
65-69	117 (16.2)	590 (15.2)	5.6 (3.8 – 6.1)
70-74	137 (18.9)	635 (16.3)	6.3 (4.6 – 6.9)
75-79	137 (18.9)	719 (18.5)	7.0 (5.6 – 7.4)
80-84	119 (16.4)	717 (18.4)	2.4 (1.3 – 3.2)
≥85	127 (17.5)	784 (20.1)	4.5 (2.5 – 5.4)
Service admitted to:			
Surgery	126 (17.4)	438 (11.3)	8.4 (6.5 – 9.2)
Internal medicine	598 (82.6)	3453 (88.7)	4.7 (4.1 – 5.3)
Appropriateness admission			
Yes	713 (98.5)	3856 (99.1)	5.0 (4.3 – 6.4)
No	11 (1.5)	35 (0.9)	17.1 (6.3 – 23.2)
Length of stay (days)			
1-2	213 (29.4)	320 (8.2)	2.2 (1.4 – 2.8)
3-4	200 (27.6)	687 (17.7)	5.7 (4.8 – 6.5)
5-6	131 (18.1)	717 (18.4)	4.0 (3.4 – 4.4)
7-8	69 (9.5)	505 (13.0)	4.0 (3.2 – 4.7)
9-10	33 (4.6)	313 (8.0)	4.8 (4.0 – 5.5)
≥11	78 (10.8)	1349 (34.7)	6.5 (6.1 – 6.9)
Number of comorbidities conditions			
0	125 (17.3)	408 (12.5)	5.3 (4.7 – 5.8)
1	71 (9.8)	385 (9.9)	9.4 (8.1 – 11.3)
2	169 (23.3)	938 (24.1)	5.1 (4.3 – 5.8)
3	174 (24.0)	992 (25.5)	3.1 (2.8 – 3.6)
≥4	185 (25.6)	1088 (28.0)	5.2 (4.9 – 6.3)

Table 2 shows the mean cost of hospital stay according to the sample's characteristics. When a comparison was made between the total cost means of appropriate and inappropriate stays, we noticed that there was no significant statistical difference in 70–74 year olds, 80–84 year olds and 85 or more year olds. Likewise, no significant differences were found between cost means of appropriate and inappropriate stays of elderly patients who were admitted for surgery services. Regarding appropriate admissions of elderly patients, a difference between means (p < 0.001) was found.

Table 2. Cost of hospital stay in the studied elderly patient sample

| | Cost of hospital stay | | | | |
| | Appropriate | | Inappropriate | | |
	Mean	(95% CI)	Mean	(95% CI)	p
Sex					
Female	1,555.6	(319.4 – 5,413.1)	2,528.0	(665.7 – 5,817.9)	p < 0.001
Male	1,432.5	(328.4 – 4,459.5)	2,132.0	(470.6 – 6,012.7)	p < 0.001
Age (years)					
60–64	1,409.4	(317.2 – 3,762.7)	2,183.2	(1,031.0 – 4,789.2)	p = 0.018
65–69	1,261.6	(314.3 – 3,105.3)	3,374.0	(481.0 – 16,991.7)	p = 0.001
70–74	1,397.5	(336.6 – 4,234.4)	1,703.3	(475.8 – 3,582.2)	p = 0.239
75–79	1,504.2	(327.5 – 5,658.4)	2,392.7	(398.0 – 7,413.3)	p = 0.010
80–84	1,679.5	(299.8 – 5,665.7)	2,395.0	(835.4 – 5,451.9)	p = 0.116
≥ 85	1,696.3	(394.6 – 4,618.6)	2,283.6	(1,002.3 – 3,887.6)	p = 0.062
Service admitted to:					
Surgery	1,191.8	(334.3 – 5,059.3)	1,611.3	(362.1 – 3,596.7)	p = 0.197
Internal medicine	1,536.0	(317.8 – 4,871.3)	2,456.2	(655.0 – 6,594.5)	p < 0.001
Appropriateness admission					
Yes	1,497.2	(323.2 – 4,931.4)	2,278.0	(763.6 – 5,942.1)	p < 0.001
No	NA		2,515.3	(363.7 – 14,772.5)	NA
Length of stay (days)					
1–2	683.0	(202.4 – 1,623.6)	365.8	(355.9 – 375.6)	p = 0.179
3–4	1,016.3	(512.2 – 1,978.7)	1,239.5	(476.1 – 2,855.7)	p = 0.020
5–6	1,548.3	(835.4 – 2,466.1)	1,403.0	(939.8 – 2,169.4)	p = 0.210
7–8	2,019.1	(1,195.4 – 2,991.3)	1,867.8	(1,181.6 – 3,145.1)	p = 0.350
9–10	2,351.6	(1,546.6 – 3,197.5)	2,398.4	(1,354.1 – 3,171.2)	p = 0.813
≥ 11	4,141.7	(2,055.5 – 7,246.8)	4,490.7	(1,935.5 – 14,042.9)	p = 0.562
Number of comorbidities conditions					
0	1,486.8	(320.6 – 4,883.8)	2,266.0	(585.1 – 6,778.6)	p = 0.038
1	1,369.7	(334.3 – 3,002.8)	2,425.8	(1,052.4 – 3,605.6)	p < 0.001
2	1,413.8	(332.2 – 4,115.5)	1,943.6	(684.3 – 4,110.3)	p = 0.010
3	1,659.6	(346.3 – 6,202.7)	2,900.7	(576.6 – 12,091.9)	p = 0.005
≥ 4	1,477.1	(282.4 – 4,272.3)	2,051.4	(600.7 – 5,048.5)	p = 0.106

NA = Not Apply

On the subject of patient days, those who remained hospitalized for three to four days showed a significant difference (p = 0.020) according to the appropriate-stay criterion. In elderly patients who showed four or mores comorbidities conditions, no significant differences were found.

The multivariate estimation by the Ordinary Least Squares model showed heterocedasticity problems (Breusch-Pagan and White tests with p < 0.001), therefore Generalized Least Squares were applied. The latter involves a regression model that is worthy to identify significant associated factors of hospitalization costs. As a dependent variable the hospitalization costs (previously transformed using the logarithm method) was used and on the right side, independent variables included in the regression model were: sex, age, inappropriate admission (0 = appropriate, 1 = inappropriate), total length of stay (appropriate + inappropriate days), comorbidities and the number of inappropriate days only. Results are shown in Table 3. The significant variables (p < 0.05) obtained through this model were, the intercept, age, inappropriate-patient days and the total number of patient days. As was expected, the latter had the greatest impact on the hospital-stay costs. Therefore, the multivariate regression showed that the hospitalization costs are highly associated with the age, total length of stay and the inappropriate length of stay.

Table 3. Generalized Least Square model using as dependent variable the transformed logarithm of hospital costs (n = 724)

Variable	β (SD)	p-value
Intercept	-135.0 (3.3)	<0.001
Sex	12.4 (39.5)	0.712
Age	4.9 (2.2)	0.005
Inappropriateness admission	-135.0 (223.1)	0.075
Patient days	222.2 (3.7)	<0.001
Comorbidities	7.7 (14.0)	0.708
Inappropriateness days	-39.7 (12.8)	0.001

β = Parameter estimate for each variable considered in the GLS model
SD = Standard Deviation

Discussion

This study reported a 12% rate of inappropriate hospitalization use, which falls within the 5% to 74% [18,20-26] ranges reported in literature.

It is worth to mention that this study has methodological weaknesses derived from the retrospective review of the clinical files and from the quality of its design; consequently it is possible that insufficient or incomplete information could generates estimate biases. Overall resource use could be underestimated and consequently, hospital stays costs could be affected [27,28].

On the other hand the study presents a selection bias related to the design of the study. Due to feasibility, those who were not admitted to the hospital were not included. Consequently the rate of false negative could be underestimated. A second stage of this study could be to test the AEP criteria in the emergencies room [16].

In spite of its weaknesses, this study is not based on certain assumptions, such as that the assistance services are always appropriate from the viewpoint of the specialist doctor who renders them, or that some socio-economic factors and clinical circumstances affect the illness' evolution, justifying hospital admission [16,27]. To overcome these circumstances, the AEP instrument was used to collect the information. This instrument is characterized by its high reliability and adequate validity to identify inappropriate hospital use [16,17].

In our study, the cause to classify hospital stays as inappropriate were attributable to the specialist doctor and/or organizational type, and could be solved by implementing interventions in the hospital assistance process, as reported in Sweden. In a study carried out at the internal medicine services of a university hospital, 15% of hospital stays were considered as inappropriate. In response to this, an intervention was implemented in the hospital assistance process resulting in a reduction of up to 9% of inappropriate stays [29].

Nevertheless, an important element to consider in the design of any strategy of this type is that patient days necessary for the recovery of the patient are difficult to determine. We must bear in mind that response to treatment is different in each individual and is frequently conditioned by the severity of the illness and its evolution time (which to begin with, led to the problem that caused the hospitalization), the patient's age (which in a certain way represents the individual's biological reserve), the presence and severity of comorbidities, timely medical assistance, psychological and emotional condition of the patient, as well as support of the social networks the patient has available [30,31].

In our study it was possible to identify that elderly patients admitted inappropriately had more inappropriate patient days (17.1%), in comparison to those whose admissions were appropriate (5.0%). Regarding costs, inappropriate stays were 55.2% higher than the appropriate stays. The frequency of inappropriate admissions was lower than we expected. This result can be due to the fact that a hospital with a high demand for services and with low resources, such as the hospital in which the study was carried out, would render evaluation of the pertinence of the hospital admission of an individual more rigorous. This was not a private hospital and/or one with private medical insurance so it strengthens the possibility that the hospital admissions and stays increases in those who really do need it. However, the study presents a selection bias related to the design of the study. Due to feasibility, those who were not admitted to the hospital were not included and the rate of false negative could be underestimated. Consequently our study included only appropriately and inappropriately admitted patients. A second stage of this study must be to evaluate the AEP criteria in the emergencies room to be able to evaluate patients inappropriate and appropriate non-admitted patients to the hospitalization areas.

Heterocedasticity problems of the Ordinary Least Squares regression were mainly due to the heterogeneous characteristics of the patients and the diversity of resources used when the patient days number increased. The dependent variable was transformed applying the natural logarithm, process which did not stabilize the variance. Accordingly, a Generalized Least Squares model was used to correct said heterocedasticity problems. The results show that age, number of patient days and inappropriate patient days are variables that have an effect on hospital-stay costs. Hence it was identified that every year of the patient's life tends to increase total hospitalization costs to US$4.9, while every appropriate patient day raises them close to US$222.2.

On the other hand, the daily cost of inappropriate hospital-stay increases the total hospitalization cost to US$182.3. This states that daily inappropriate costs are 18% under daily appropriate costs due a lower intensive medical care. Therefore, this inappropriate patient day cost reflects the fact that, in view of a lack of an additional diagnostic and/or treatment plan in the clinical file, all the resources for patient assistance are not used as in the case of appropriate hospital-stay. This includes also other resources such as laboratory tests and office examinations, as well as inter-visits to other specialists.

This evaluation identifies that there is an inappropriate and unnecessary use of hospital service resources provided to the elderly patients. In other words, a variety of human and technical resources and infrastructure are used in circumstances were they are not indicated form a strictly medical viewpoint, leading to a raise of hospital-stay costs. Authors recommend that in the future this type of studies should be carried out for elderly patients with specific diseases.

The elderly population has been identified as the greatest consumer of health services, generating expenses which mainly arise from their hospitalization in medical centers [32]. Health Systems around the world are facing very similar challenges. Particularly in Latin America, where the demographic transition has been too fast, the increasing demand of health services by the elderly is a common issue.

It is well known that hospital resources are not used in a proper way. In some occasions, elderly patients receive services that they do not really need. Others, benefits received are not significant or the care given at the hospital could be given through other health care schemes less expensive and more appropriate such as nurses at home, day care hospitals and so on [33,34]. In that sense, our results are of interest for countries with similar social security systems. Mexico, as many other developing countries, has to move on to different organization models. Consequently, even when costs obviously can not be generalized, the whole sense of the paper can be of important for other regions.

Therefore, an improvement of elderly patient's assistance and an optimum utilization of medical assistance resources, especially of the hospitalization services, represent an important challenge for health service providers worldwide.

Conclusion

Elderly patients who were inappropriately hospitalized had a higher rate of inappropriate patient days. The average of inappropriate patient days cost is considerably higher than appropriate days. In this study, inappropriate hospital-stay causes could be attributable to physicians and current organizational management.

Competing Interests

The authors declare that they have no competing interests.

Authors' Contributions

JFM-Q originated the idea for this study, did the research proposal, data analysis, and prepared the manuscript. CG-P, IC-H, and TJ-C contributed to the research proposal, reviewed the analysis, and participated in the preparation of the manuscript. CE-B participated in the interpretation of the data and in the discussion of the paper. GM-C participated in the research proposal and reviewed the manuscript. SS-G designed and conducted the original proposal and was involved in the data analysis and in the preparation and discussion of the manuscript. All authors read and approved the final manuscript.

Acknowledgements

The authors are grateful to the personnel from the Mexican Social Security Institute (IMSS), "Dr. Carlos Mac Gregor Sánchez Navarro" Zone 1 Regional Hospital in Mexico City. This project was sponsored by the IMSS's Fund for Promotion of the Health Research: FOFOI-IMSS2004/105 grant, and the CONACyT's Sectorial Fund for Health Research and Social Security: Salud-2004-C01-23 grant.

References

1. Envejecimiento de la población en México: reto del siglo XXI [http://www.conapo.gob.mx/ index.php?option=com_content&view=article&id=340&Itemid=15] CONAPO, México; 2004.

2. Lorenzo S: Revisión de utilización de recursos. Estudios realizados en España. Rev Calidad Asistencial 1997, 12:140–146.

3. Peiró Moreno S, Portella E: Identification of the inappropriate use of hospitalization: the search for efficiency. Med Clin (Barc) 1994, 103:65–71.

4. Anton P, Peiro S, Aranaz JM, Calpena R, Compan A, Leutscher E, Ruiz V: Effectiveness of a physician-oriented feedback intervention on inappropriate hospital stays. J Epidemiol Community Health 2007, 61:128–34.

5. Palmesano-Mills C: Common problems in hospitalized older adults. J Gerontol Nurs 2007, 33:48–54.

6. Navarro G, Prat-Marín A, Asenjo M, Menacho A, Trilla A, Salleras L: Review of the utilisation of a university hospital in Barcelona (Spain): evolution 1992–1996. Eur J Epidemiol 2001, 17:679–84.

7. Zambrana García JL, Delgado Fernández M, Cruz Caparrós G, Díez García F, Martín Escalante MD, Salas Coronas J: Factors associated with inappropriate hospitalization at an internal medicine department. Med Clin (Barc) 2001, 116:652–654.

8. Luquero Alcalde FJ, Santos Sanz S, Pérez Rubio A, Tamames Gómez S, Cantón Alvarez MB, Castrodeza Sanz J: Factors determining inappropriate days of stay in a third-level hospital. Gac Sanit 2008, 22:48–51.

9. Polder JJ, Bonneux L, Meerding WJ, Maas PJ: Age-specific increases in health care costs. Eur J Public Health 2002, 12:57–62.

10. Polder JJ, Barendregt JJ, van Oers H: Health care costs in the last year of life – the Dutch experience. Soc Sci Med 2006, 63:1720–1731.

11. McKay NL, Deily ME: Cost inefficiency and hospital health outcomes. Health Econ 2008, 17:833–848.

12. Sangha O, Schneeweiss S, Wildner M, Cook EF, Brennan TA, Witte J, Liang MH: Metric properties of the appropriateness evaluation protocol and predictors of inappropriate hospital use in Germany: an approach using longitudinal patient data. Int J Qual Health Care 2002, 14:483–492.

13. Hammond CL, Pinnington LL, Phillips MF: A qualitative examination of inappropriate hospital admissions and lengths of stay. BMC Health Serv Res 2009, 9:44.

14. Drummond MF, Sculpher MJ, Torrance GW, O'Brien BJ, Stoddart GL: Methods for the Economic Evaluation of Health Care Programs. Third edition. Oxford: Oxford University Press; 2005.

15. Walley T, Haycox A, Boland A: Pharmacoeconomics. London: Churchill Livingstone; 2003.

16. Sánchez-García S, Juárez-Cedillo T, Mould-Quevedo JF, García-González JJ, Contreras-Hernández I, Espinel-Bermudez MC, Hernández-Hernández DM, Garduño-Espinosa J, García-Peña C: The hospital appropriateness evaluation protocol in elderly patients: a technique to evaluate admission and hospital stay. Scand J Caring Sci 2008, 22:306–313.

17. Peiró S, Meneu R, Roselló-Pérez M, Portella E, Carbonell-Sanchís R, Fernández C, Lázaro G, Llorens MA, Martínez-Mas E, Moreno E, Ruano M, Rincón A, Vila M: Validity of the hospitalization Appropriateness Evaluation Protocol. Med Clin (Barc) 1996, 107:124–129.

18. Booth BM, Ludke RL, Wakefield DS, Kern DC, Burmeister LF, Fisher EM, et al.: Nonacute days of care within department of veterans affairs medical centers. Med Care 1991, (suppl 29):AS51-AS63.

19. Restuccia JD, Payne S, Lenhart G, Fulton J: Assessing the appropriateness of hospital utilization to improve efficiency and competitive position. Health Care Manage Rev 1987, 12:17–27.

20. Bare ML, Prat A, Lledo L, Asenjo MA, Salleras L: Appropriateness of admissions and hospitalization days in an acute-care teaching hospital. Rev Epidemiol Sante Publique 1995, 43:328–336.

21. Santos-Eggiman B, Paccaud F, Blanc T: Appropriateness of utilization: an overview of the Swiss experience. Int J Qual Health Care 1995, 7:227–232.

22. Fellin G, Apolone G, Tampieri A, Bevilacqua L, Meregalli G, Minella C, Liberati A: Appropriateness of hospital use: an overview of Italian studies. Int J Qual Health Care 1995, 7:219–225.

23. Bentes M, Gonsalves ML, Santos M, Pina E: Design and development of a utilization review program in Portugal. Int J Qual Health Care 1995, 7:201–212.

24. Lorenzo S, Sunol R: An overview of Spanish studies on appropriateness of hospital use. Int J Qual Health Care 1995, 7:213–218.

25. Hwang JI: Characteristics of patient and healthcare service utilization associated with inappropriate hospitalization days. J Adv Nurs 2007, 60:654–662.

26. Ingold BB, Yersin B, Wietlisbach V, Burckhardt P, Bumand B, Büla CJ: Characteristics associated with inappropriate hospital use in elderly patients admitted to a general internal medicine service. Aging 2000, 12:430–438.

27. Suárez García F, Oterino de la Fuente D, Peiró S, Librero J, Barrero Raya C, Parras García de León N, Crespo Pérez MA, Peréz-Martín A: Factors associated with the use and adaptation of hospitalization in people over than 64 years of age. Rev Esp Salud Publica 2001, 75:237–48.

28. Ramos-Cuadra A, Marión-Buen J, García-Martín M, Fernández Gracia J, Morata-Céspedes MC, Martín-Moreno L, et al.: Use of the appropriateness evaluation protocol: the role of medical record quality. The effect of completeness of medical records on the determination of appropriateness of hospital days. Int J Qual Health Care 1995, 7:267–275.

29. Kossovsky MP, Chopard P, Bolla F, Sarasin FP, Louis-Simonet M, Allaz AF, Perneger TV, Gaspoz JM: Evaluation of quality improvement interventions to reduce inappropriate hospital use. Int J Qua Health Care 2002, 14:227–232.

30. Perlado F: Capítulo 5 "Geriatría Hospitalaria" en Teoría y Práctica de la Geriatría. In 1a Edic. Madrid: Ediciones Díaz de Santos; 1995.

31. Foucault M: Capítulo 5 "La lección de los hospitales" en El nacimiento de la clínica. Una arqueología de la mirada médica. Ciudad de México: Editorial Siglo XXI 1996.

32. Peleg R, Press Y, Asher M, Pugachev T, Glicensztain H, Lederman M, Biderman A: An intervention program to reduce the number of hospitalizations of elderly patients in a primary care clinic. BMC Health Serv Res 2008, 8:36.

33. Elkan R, Kendrick D, Dewey M, Hewitt M, Robinson J, Blair M, Williams D, Brummell K: Effectiveness of home based support for older people: systematic review and meta-analysis. BMJ 2001, 323:719–725.

34. Steeman E, Moons P, Milisen K, De Bal N, De Geest S, De Froidmont C, Tellier V, Gosset C, Abraham I: Implementation of discharge management for geriatric patients at risk of readmission or institutionalization. Int J Qual Health Care 2006, 18:352–358.

The Use of Edinburgh Postnatal Depression Scale to Identify Postnatal Depression Symptoms at Well Child Visit

Vincenzo Currò, Emilia De Rosa, Silvia Maulucci,
Maria Lucia Maulucci, Maria Teresa Silvestri,
Annaluce Zambrano and Vincenza Regine

ABSTRACT

Objectives

1) to evaluate the role of the pediatrician in detecting postnatal depression (PD) symptoms by the Edinburgh Postnatal Depression Scale (EPDS); 2) to detect factors increasing the risk of PD and, 3) to assess the importance of scores gained from fathers' questionnaire.

Methods

We surveyed 1122 mothers and 499 fathers who were assessed using the EPDS during the first well-child visit. After 5 weeks, high scoring parents, completed a second EPDS. High scoring parents were examined by a psychiatrist who had to confirm the PD diagnosis.

Results

26.6% of mothers and 12.6% of fathers at the first visit, 19.0% of mothers and 9.1% of fathers at the second visit, gained scores signaling the risk of PD. Four mothers and two fathers had confirmed PD diagnosis. Younger maternal age, non-Italian nationality and low socio-economic condition were related to higher EPDS scores.

Conclusion

PD is common in the average population. Using a simple and standardized instrument, pediatricians are able to detect parents with higher risk of suffering from PD.

Background

Postnatal depression (PD) is the most common disorder following childbirth and a social problem for public welfare: ten to fifteen women out of one hundred suffer from this disorder [1]. PD is a severe condition that has been described as "a thief who steals maternity"; up to 50% of the cases are diagnosed, and approximately 49% of women who seek help feel desperately depressed [2]. If untreated, a large number of these mothers continue to be depressed until the end of the first and the second postnatal years.

Women who suffer from PD are exposed to an increased risk of future depression, relapses, thoughts of abusing their children, and face difficulties in the child-mother relationship [3]. Maternal Depression (MD) may have a strong negative impact on the social, cognitive and behavioral development of children, including an increased rate of behavioral problems at school [4,5]. The obstetrician should be the first to identify those mothers who risk developing depression, but this can be quite difficult as symptoms often appear after the routine 4 to 6 week postnatal examination. Moreover, mothers with PD often do not recognize the symptoms of depression. This is a result of the difficulty in identifying the following signs as symptoms of PD: weight loss, irritability, crying fits and fatigue (often considered as the physiological adaptation to life with a newborn child). This is the reason why many women do not receive an immediate diagnosis or an appropriate treatment program [6]. Pediatricians may be the only medical workers that are routinely met by mothers during the first twelve months of the baby's life [7]. A self- report rating scale routinely administered in pediatrics could be a useful tool to recognize the risk of PD. The Edinburgh Postnatal Depression Scale is the most widely used screening scale for PD [8]. It has been validated in Holland, Australia, Portugal, Sweden, Italy, Spain, United Arab Emirates, France and India [6,8-15].

The EPDS has already been administered in a pediatrics setting at the University Rochester Medical Centre (NY), although the impact of this screening instrument on the visit time was not assessed. However, the length of the visit is an important factor to take into consideration when working in a busy pediatric clinic. As a matter of fact, if we extend visit times, we risk reducing the total number of daily visits.

In this research project we focused on PD risk in mothers who have recently given birth. Many studies highlighted that the most significant risk factors for PD are: young maternal age, absence of a social support, immigration, and lack of a supporting spouse [16]. In addition, we also took into account what the outcome would be if new fathers were evaluated as well. To date, only a few studies have investigated PD risk in fathers, and no study has been conducted in Italy. Men who have manual or working class occupations and low social integration are more likely to become depressed [9]. In this study we will refer to parents at both low and high risk of suffering from depression on the basis of the EPDS score. In fact, a high EPDS score does not mean postnatal depression but only a high risk of suffering from PD. A psychiatrist was involved to confirm the diagnosis of PD. The aims of our study were: 1) to check the feasibility of assessing the risk of PD in parents using the EPDS; 2) to provide correlations between PD risk and socio-demographic information, both in mothers and in fathers; 3) to check correlations between high EPDS scoring fathers in couples with EPDS positive mothers and high EPDS scoring fathers in couples with EPDS negative mothers.

Methods

Setting

The study was conducted at the Pediatric Clinic of the Policlinico A. Gemelli, Catholic University Hospital, Rome.

This clinic admits 4000 patients a year, and it is a teaching site for pediatric residents and medical students.

Participants

Our team included: a senior pediatrician, two pediatric residents, two psychologists, and two psychiatrists. The EPDS was proposed to all parents, regardless of age and nationality, at the first postnatal check-up within the first year of baby's birth. Unmarried women were also included. Postnatal check-up examinations were carried out by the senior pediatrician with the aid of pediatric residents.

Research Tools

EPDS is a paper-and-pencil self-reporting questionnaire composed of 10 questions and a 0-3 point scale. The cut-off score is 9 for women and 7 for men [11,17,18].

Procedure

EPDS was proposed by a pediatrician, before clinical examination. The informed consent was obtained by explaining the meaning of the EPDS and we asked participants to complete questionnaires, without any help, and to answer according to their feelings during the previous seven days. In the case of high scores, the EPDS was repeated after five weeks, especially when we visited babies in the first 15 days of life, as, after delivery, many women experience 'baby blues.' Baby blues is considered a normal stage of early motherhood, usually disappearing some days after delivery.

The EPDS was translated and validated into several languages (Italian, French, Spanish, English, Arabic, and Punjabi) for parents who did not understand Italian. We translated the EPDS into Singhalese and used it, although this version was not officially validated.

Whenever our test results were rated as 'high,' we told mothers or fathers or sometimes both parents, to undergo a psychologist-psychiatrist examination. Without an appropriate psychiatric evaluation, a high EPDS score does not mean PD. Sometimes, in the self-reporting test, people report anxiety, mood instability, depressed mood, and other transient emotional disturbances which disappear within a few hours or days. A PD diagnosis requires a woman to be experiencing dysphoric mood and several other symptoms such as appetite, sleep or psychomotor disturbances, excessive feeling of guilt, fatigue and suicidal thoughts for a minimum of two weeks.

Study Population

This cross-sectional study started in January 2005 and ended in November of the same year. 1130 infants were examined at the first postnatal check-up (median 17 days, range 15-20 days). We proposed the EPDS to 1628 parents; 1621 subjects, 1122 mothers of the 1127 (99.6%) and 499 fathers of the 501 (99.6%) completed the EPDS and were included in this study. Five mothers with previous depression symptoms and two fathers who did not complete the EPDS were excluded. The male group was smaller than the female one, due to the fact that fathers were investigated only from August onwards and mothers often attended the clinic alone.

We excluded mothers with a history of depression since we intended to check only symptoms of PD, which is a perinatal pathology, not to be confused with depression.

Statistical Analysis

To evaluate the feasibility of assessing postnatal depression symptoms, we calculated by how many minutes (average and range) the completion of the questionnaire extended the visit time, and the percentage of compilers.

The prevalence of the risk of PD was calculated as the percentage of mothers who scored ≥ 10 and as the percentage of fathers who scored ≥ 8. The characteristics of subjects at high PD risk and at low PD risk were compared using the Chi-squared test for each following characteristic: maternal and paternal age, marital status, employment, educational level, nationality, nursing, number of pregnancies, gestational age, delivery, mother's and father's pathologies, baby's weight at birth, gender, infant hospitalization and pathologies, and the season of the interview. Crude odds ratios (OR) were also calculated for all variables and adjusted OR were also calculated for variables where the univariate analysis showed a statistically significant association (p-value for Chi squared test < 0.05). Adjusted OR were calculated with the construction of a multivariate logistic regression model using the backward elimination method. The fit of the model was assessed using the Hosmer-Lemenshow test.

Results

Length of the Visit

The time range taken to complete the test was 2-7 minutes, with a mean of 3.28 minutes for women and 3.22 minutes for men. Less time was employed if parents had no problems with the language: 2.98 minutes (range: 2-4) for Italian females and 2.98 minutes (range: 2-5) for Italian males; 5.1 minutes (range: 2-7) and 5.6 minutes (range: 3-7) for foreign females and males, respectively. For foreigners who did not have the text translated into their mother tongue, the time was longer: 6.00 minutes (range: 5-7) for both females and males. For foreigners with the test translated into their mother tongue, the time was 3.73 minutes (range: 2-6) for females and 4.33 minutes (range: 3-5) for males.

Socio-Demographic Characteristics of the Sample

Newborns

Table 1 shows the demographic and clinical characteristics of the 1,130 infants: 51.1% were male, 79.8% were Italian, 62.6% were first-born, 25.3% had pathologies and 2.1% were hospitalized.

Table 1. Demographic and clinical characteristics of 1,130 infants

	Number	Percentage
Sex		
Male	577	51.1
Female	553	48.9
Nationality*		
Italian	896	79.8
Non Italian	227	20.2
Brothers		
No	707	62.6
Yes	423	37.4
Twin birth		
No	1127	99.7
Yes	3	0.3
Weight at birth*		
≤ 2500 g	64	5.7
> 2500 g	1064	94.3
Gestational Age*		
≤ 36 wk	74	6.6
> 36 wk	1052	93.4
Nursing*		
Breastfeeding	756	69
Bottlefeeding	115	10.5
Both	224	20.5
Hospitalization		
No	1106	97.9
Yes	24	2.1
Pathology		
No	844	74.7
Yes	286	25.3

*information is not available for all infants

As "pathology," we considered any problems, even not serious, signed in the discharge papers of the newborn babies that could create anxiety to parents (e.g. jaundice, hip instability, cefaloematoma, hypocalcemia, patent foramen ovale).

Hospitalization regarded surgery, phototherapy, infections, prematurity, respiratory distress syndrome (RDS).

Mothers

The mean age of the 1,122 mothers assessed was 32.9 years (SD 4.9), 62.7% were primiparae with a mean age of 31.7 years, 96.9% were married or lived with their partner and 20.2% were non-Italian. 85.1% had a school diploma or degree, 72.9% were employed; 713 women (63.6%) had natural delivery, 409 (36.4%) had a caesarean delivery, and 49 women had previous spontaneous abortion.

Countries of origin of the 223 non-Italian mothers were: 43.5% East Europe, 25.7% South America, 14.1% Asia, 10.5% West Europe and North America, and 6.2% Africa.

Fathers

The mean age of the 499 fathers assessed was 36.3 years (SD 5.5); 99.6% were married or lived with their partner; 10.1% were non-Italian; 79.7% had a school diploma or degree and 98.6% were employed.

Countries of origin of the 50 non-Italian fathers were: 32.0% East Europe, 24.0% South America, 20.0% Asia, 12% West Europe and North America, 12.0% Africa.

Morbidity

Mothers' Morbidity

Interviewed mothers had an EPDS mean score of 7.11 (SD 4.42). 298 (26.6%) mothers had an EPDS score $\geq$ 10 and 824 (73.4%) had an EPDS score <10. For those mothers with a low depression risk the mean score was 4.98 (SD 2.55), whereas for mothers with a high risk of depression, the mean score was 13.01 (SD 2.88).

Fathers' Morbidity

The total EPDS mean score was 3.86 (SD 3.12). 63 fathers had an EPDS score $\geq$ 8 with a 12.6% risk of depression. The mean score for 63 high risk fathers was 9.98 (SD 2.11) and 2.97 (SD 2.07) for 436 low risk fathers.

Correlations between the Risk of Pd and Socio-Demographic Information

Mothers

Table 2 shows the demographic and clinical characteristics of both high and low depression risk mothers, the occurrence of PD risk, crude OR (COR) and adjusted OR (AOR).

Table 2. Demographic and clinical characteristics of high PD risk mothers vs low PD risk mothers

	Mothers at high risk of PD (n = 298)	Mothers at low risk of PD (n = 824)	Prevalence %	p-value**	Crude OR (95% CI)	Adjusted OR (95% CI)
Age group[a]						
Mean age (SD)	31.95 (5.60)	33.23 (4.54)		0		
< 25 years	35	25	58.3		4.25 (2.50-7.23)	3.12 (1.88-5.82)
≥ 25 years	263	798	24.8		1	1
Nationality[a]						
Italian	205	687	23	0	1	1
Non Italian	92	131	41.3		2.35 (1.73-3.20)	2.01 (1.43-2.82)
Marital status						
Married^	283	805	26	0.019	1	1
Single	15	19	44.1		2.25 (1.13-4.48)	1.49 (0.69-3.19)
Educational level[a]						
Primary	6	4	60	0.001	4.58 (1.28-16.38)	2.30 (0.58-9.12)
Middle	55	98	35.9		1.71 (1.19-2.46)	1.34 (0.90-1.99)
High^	234	715	24.7		1	1
Number of pregnancies[a]						
1	209	494	29.7	0.002	1.56 (1.17-2.07)	1.33 (0.98-1.80)
≥ 2	89	328	21.3		1	1
Feeding of infant[a]						
Breastfeeding	178	573	23.7	0.003	1	1
Bottlefeeding[a]	109	227	32.4		1.54 (1.16-2.04)	1.67 (1.20-2.17)
Employment[a]						
No	94	209	31	0.097	1.28 (0.95-1.72)	
Yes	213	605	26		1	
Pathology of mother						
No	164	473	25.7	0.479	1	
Yes	134	351	27.6		1.10 (0.84-1.44)	
Pathology of father[a]						
No	85	250	25.4	0.988	1	
Yes	41	121	25.3		1.00 (0.63-1.57)	
Gestational Age[a]						
≤ 36 wk	19	55	25.7	0.872	0.96 (0.54-1.69)	
> 36 wk	277	767	26.5		1	
Delivery[a]						
Vaginal	173	535	24.4	0.058	1	
Abdominal	120	285	29.6		1.30 (0.99-1.71)	
Gender of infant						
Male	149	421	26.1	0.747	1	
Female	149	403	27		1.04 (0.79-1.37)	
Weight at birth[a]						
≤ 2500 g	19	45	29.7	0.543	1.19 (0.68-2.06)	
> 2500 g	277	779	26.2		1	

Table 2. *(Continued)*

Pathology of infant					
No	217	621	25.9	0.687	1
Yes	77	207	27.2		1.06 (0.78-1.46)
Hospitalization of infant					
No	286	814	26	0.893	1
Yes	6	16	28.6		1.07 (0.37-2.93)
Season of interview*					
Autumn	53	176	23.1	0.568	1
Winter	57	164	25.8		1.15 (0.75-1.77)
Spring	100	254	28.2		1.31 (0.89-1.92)
Summer	86	230	27.2		1.24 (0.84-1.84)

Note: Variables with a p-value < 0.05 were included in the multivariate analysis
*Information is not available for all mothers
** p-value for Chi-square test
^Include common-law wife
^Include diploma and degree
^Include mixed breastfeeding

The occurrence of a high risk of PD is significantly higher (p < 0.05) among <25 years of age vs. ≥ 25 years of age (58.3% vs. 24.8%); among non-Italians vs. Italians (41.3% vs. 23.0%); among singles compared to married women (44.1% vs. 26.0%); among those with a lower level of education (60.0% and 35.9%) compared to those with a higher level of education (24.7%); among primiparae vs. multiparae (29.7% vs. 21.3%) and among bottle-feeding mothers vs. breast-feeding mothers (32.4% vs. 23.7%). The multivariate logistical regression model shows that mothers with high depression risk were: <25 years of age (AOR, Adjusted Odds Ratio, 3.12; 95%; CI 1.88-5.82), non-Italian (AOR 2.01, 95%; CI 1.43-2.82), bottle-feeding (bottle-feeding vs. breast-feeding AOR = 1.67; CI 1.20-2.17). Marital status and educational level which appeared to be significant in the univariate analysis were no longer significant in the multivariate analysis. We did not find any significant association (p-value for Chi squared test p > 0.05) between PD risk and low birth weight babies, gender, newborn pathologies and hospitalization, gestational age, delivery, employment, maternal and paternal pathology, and seasonality. There were no significant differences among non-Italian mothers (p > 0.05): the occurrence of a high risk of PD was 45.8% for mothers born in East Europe, 36.7% in South America, 42.3% in Asia, and 58.3% in Africa.

Fathers

Table 3 shows the demographic and clinical characteristics of both high risk and low risk fathers, the prevalence of father risk of suffering from depression, COR and AOR. The occurrence of the risk of paternal depression was significantly higher (p < 0.01) among <30 years of age vs. ≥ 30 years of age (27.3% vs. 11.2%); among non-Italians vs. Italians (24.0% vs. 11.3%); among those with a lower level of education (primary or middle education 21.8% vs. high education 10.3%); among unemployed or unskilled employment vs. other employment (30.2% vs. 10.4%).

Table 3. Demographic and clinical characteristics of high PD risk fathers vs low PD risk fathers

	Fathers at high risk of PD (n = 63)	Fathers at low risk of PD (n = 436)	Prevalence %	p-value[oo]	Crude OR (95% CI)	Adjusted OR (95% CI)
Age group						
Mean age (SD)	35.32 (6.72)	36.41 (5.34)		0.002		
< 30 years	12	32	27.3		2.97 (1.44-6.13)	1.98 (0.86-4.61)
≥ 30 years	51	404	11.2		1	1
Nationality[o]						
Italian	50	394	11.3	0.01	1	1
Non Italian	12	38	24		2.49 (1.22-5.07)	1.23 (0.51-2.95)
Educational level[o]						
Middle[o]	22	79	21.8	0.002	2.42 (1.36-4.29)	1.71 (0.92-3.20)
High[oo]	41	354	10.3		1	1
Employment[o]						
Unskilled[o]	16	37	30.2	0	3.73 (1.93-7.23)	2.24 (1.02-4.96)
Skilled	46	397	10.4		1	1
Marital status[o]						
Married^	62	434	12.5	0.111	1	
Single	1	1	50		7.0 (0.43-113.35)	
Pathology of father						
No	42	293	12.5	0.933	1	
Yes	21	143	12.8		1.02 (0.58-1.79)	
Pathology of mother						
No	43	259	14.2	0.179	1	
Yes	20	177	10.2		0.68 (0.37-1.24)	
Number of sons[o]						
1	41	272	13.1	0.745	1.10 (0.61-1.98)	
≥ 2	22	160	12.1		1	
Gestational Age						
≤ 36 wk	4	25	13.8	0.845	1.11 (0.37-3.32)	
> 36 wk	59	411	12.6		1	
Delivery[o]						
Vaginal	39	281	12.2	0.777	1	
Abdominal	23	153	13.1		1.08 (0.62-1.88)	
Gender of Infant						
Male	37	221	14.3	0.232	1	
Female	26	215	10.8		0.72 (0.41-1.27)	
Weight at birth						
≤ 2500 g	4	24	14.3	0.785	1.16 (0.39-3.47)	
> 2500 g	59	412	12.5		1	
Pathology of infant						
No	42	307	12	0.231	1	
Yes	24	126	16		1.39 (0.78-2.47)	
Hospitalisation of infant						
No	58	421	12.1	0.699	1	

Table 3. *(Continued)*

Yes	3	17	15		1.28 (0.29-4.83)
Feeding of infant[*]					
Breastfeeding	33	274	10.7	0.205	1
Bottlefeeding[#]	25	145	14.7		1.43 (0.79-2.59)
Season of interview[*]					
Autumn	25	201	11.1	0.555	1
Winter	0	7	0		-
Spring	6	32	15.8		1.51 (0.57-3.96)
Summer	31	196	13.7		1.27 (0.72-2.23)

Note: Variables with a p-value < 0.05 were included in the multivariate analysis
[*]Information is not available for all fathers
[**] p-value for Chi-square test
[+] include primary education
[***]include diploma and degree
[^]include unemployed
[^]include common-law husband
[#] include mixed breastfeeding

The multivariate logistic regression model shows that employment was only associated with depression risk among fathers; in particular, the risk of depression was double (AOR 2.24; 95% CI 1.02-4.96) among unemployed or unskilled employed fathers vs. skilled employed fathers. Age, nationality and level of education which appeared to be significant in the univariate analysis were no longer significant in the multivariate analysis. We found no statistical association (p-value for Chi squared test p > 0.05) between paternal risk of suffering from depression and the other variables considered: marital status, number of sons, low birth weight babies, gender, type of feeding, newborn pathologies and hospitalization, gestational age, delivery, maternal and paternal pathology, and seasonality.

There were no significant differences among nationalities of non-Italian fathers (p-value > 0.05): the occurrence of a high risk of PD was 12.5% for fathers born in East Europe, 33.3% in South America, 40.0% in Asia, and 33.3% in Africa.

A High Epds Score Must be Confirmed by a Psychiatrist

The psychiatrist was required to confirm the PD diagnosis in the 147 mothers and 22 fathers that attended the second visit and had scores signaling a risk of PD, 19.0% and 9.1% respectively. The psychiatric evaluation lowered the range of the EPDS positivity by 4 mothers and 2 fathers. The low number of parents with a diagnosis of PD did not permit to calculate the degree of association between the characteristics of the PD group and the factors found to be predictive of an increased risk for PD.

Couple Morbidity

497 mothers were interviewed with their partner. This sample of mothers was composed of women who were: Italian (83.5%), married (99.4%), and skilled employees (23.4%). No statistical differences were found between the occurrence of the risk of depression in mothers with or without a partner (25.3% and 27.6% respectively).

The prevalence of couples who risked suffering from depression (risk in both parents: mother with EPDS ≥ 10, father with EPDS ≥ 8) was 6.2% (31 couples). 340 couples (68.4%) did not have high scores. We observed 126 cases with the risk of depression in one parent: 95 (19.1%) in mothers and 31 (6.2%) in fathers.

The percentage of fathers with EPDS ≥ 8 was significantly higher ($p < 0.01$) among fathers in couples with high risk mothers (24.6%) than among fathers in couples with low risk mothers (8.4%). The risk of depression in fathers of couples with high risk mothers was approximately four times higher (OR 3.6; 95% CI 2.1-6.2) than in those of couples with low risk mothers.

Discussion

We carried out a pilot study in order to verify the possibility to use the EPDS and to detect its limitations. We analyzed the problems which emerged from the American study, such as the length of the visit [7]. Glaze and Cox performed a computerized version of psychiatric rating scales that may be less time-consuming than the pencil-and-paper method [19]. We have not used computerized versions of the test yet, but we are carrying out the self-reporting test during the anamnesis. Collecting data extended visit time from 1 to 2 minutes for foreign parents who did not have the EPDS translated into their native language. As to Italian parents, we did not notice any extension of the visit time. Thus, while the pediatrician collected information from papers (discharge form, nursery, vaccine book, etc.) and recorded them in a file in 5 minutes, parents could complete the test with a little effort to save time. Every time we introduce an innovative change into a well-established routine of a busy hospital we must be very careful not to increase visit times previously arranged with the directional management in order not to increase costs or cause a loss in profits. The high participation rate (99.6%) demonstrates that the EPDS is easy to understand.

Our study confirmed that the risk of PD is more common in foreign women. In London, Onozawa showed that women coming from ethnic minorities or from a non-English speaking background should be regarded as high risk group for postnatal depression [20]. Pregnancy, giving birth and bringing up a child, are also a psychological/cultural matter. We believe that environmental support and cultural models could fail in a migrant context, which is why immigrant mothers often live in a high psychic risk situation. In such critical situations we create a protective net not only medically but also socially, involving a welfare officer.

Results from our research show that fathers have lower depression rates than mothers [21,9]. However, the little number (two) of father with PD did not allow any conclusions. In our study the higher risk of paternal depression was associated with a lower level of employment. Moreover, work instability raises PD rate in fathers. Fathers give their family both a psychological and material support. If fathers do not feel capable of fulfilling this task, they could become depressed [22]. This could be one of the reasons why, in our society, males enter parenthood later than females.

In our study, the risk of PD was significantly associated with younger mothers (< 25 years of age) and the univariate analysis showed that younger fathers (< 30 years of age) were more frequently exposed to the risk of depression. Adolescents or young parents often perceive their child's birth as an obstacle to their identity [22]. Parents who are not able to fulfill their life projects or to adapt to the role of parents could become depressed. We also found that early diagnosis is important in parents with a high risk of PD. Prevention is necessary for those who are vulnerable, i.e., migrant or adolescent parents. Psychological and social support could decrease the risk of PD [23].

The risk of PD was lower in breast-feeding mothers than in bottle-feeding mothers. Can the absence of PD lead to an increased probability of breastfeeding or the breastfeeding is a protective factor against PD? [24]. Our results demonstrated an association between the two variables, independent from the other factors taken into account, and not a causal pathway in which breastfeeding "protects" against PD. The reverse can also be true: the absence of PD can lead to an increased probability of breastfeeding that perhaps is more plausible. A statistical association found in a multivariate analysis is not enough to demonstrate the presence of a causal relationship.

We found no difference in the occurrence of PD during different seasons [25].

Our results confirmed that the risk of depression for fathers in couples where the mother is depressed is approximately four times higher than that of fathers in couples with non-depressed mothers.

Limitations of this Study

A possible limitation of this study is that it was conducted in just one pediatric clinic, which was situated in an academic medical centre. Postnatal child assessment is made by family pediatricians in their private offices, where a team comprising psychologists and psychiatrists is not usually available. The extension to other practices, such as in pediatric family physician practices, would require further investigation. The feasibility of performing an evaluation with the EPDS in a well structured and trained setting does not mean that it can automatically be proposed for a more widespread and general use.

Moreover, the EPDS was used at each first well-child visit, even when the children were a few days old. It would be better to administer the EPDS from 40 days to 12 months of baby's birth, in order to check PD and overcome the risk of postnatal blues.

Another limitation is that we performed the second EPDS only on 147 mothers and 22 fathers, instead of on the 298 mothers and 63 fathers, with an high first EPDS. Therefore the high depression scores at the second visit needs to be evaluated with caution, due the high lost at the follow up. The question of whether EPDS scores or demographic characteristics at the first postnatal visit predict a diagnosis of PD would require further investigations.

Another limitation was that we did not have information regarding parent follow-up or treatment with the diagnosis of PD, thus limiting our ability to comment on the effectiveness of depression symptom detection.

We also did not examine the influence of depression on early breastfeeding termination [26]. However, breastfeeding difficulties and subsequent early breastfeeding termination may encourage postnatal depression symptoms to appear.

Conclusion

The authors presented results of a personal experience in the setting of an outpatient pediatric clinic for newborn infants at a large and specialized teaching hospital. The aim of this study was to detect factors increasing the risk of PD in the mothers and the fathers using EPDS.

Mothers and fathers often experience a moment of crisis and loneliness which they are not able to discuss because of their new frenetic lifestyle. Sometimes such problems are reported in the media as a kind of tragedy.

To assess PD in parents is very important because psychological treatments improve very quickly maternal mood and mother-infant interaction [23].

Abbreviations

EPDS: Edinburgh Postnatal Depression Scale; PD: postnatal depression; OR: odds ratios; COR: crude odds ratios; AOR: adjusted odds ratios.

Competing Interests

The authors declare that they have no competing interests.

Authors' Contributions

AZ and MTS administered the test and collected data. VR performed the statistical analysis. SM collected and interpreted data. MLM drafted and revised the manuscript. VC and EDR conceived the study, participated in its design and coordination and helped to draft the manuscript.

All authors read and approved the final manuscript.

Acknowledgements

We would like to thank Dr. Francesca Cipriani, Dr. Augusto Zani, and Dr. Riccardo Cocchi for helping in the translation.

References

1. Nonacs R, Cohen LS: Postpartum mood disorders: Diagnosis and treatment guidelines. J Clin Psychiatry 1998, 59(12):34–40.

2. Georgiopoulos A, Bryan T, Yawn B, et al.: Population-based screening for postpartum depression. Obste Gynecol 1999, 653–657.

3. Cicchetti D, Rogosch F, Toth S: Maternal depressive disorder and contextual risk: Contributions to the development of at insecurity and behavior problems in toddlerhood. Dev Psychopathol 1998, 10:283–300.

4. Cicchetti D, Rogosch F, Toth S, et al.: Affect, cognition and the emergence of self- knowledge in the toddler offspring of mothers with depression. J Exp Child Psychol 1997, 67:338–362.

5. Sinclair D, Murray L: Effects of postnatal depression on children's adjustment to school. Br J Psychiatry 1998, 72:58–63.

6. Pop VJ, Komproe IH, van Son MJ: Characteristics of the Edinburgh Postnatal Depression Scale in the Netherlands. J Affect Disord 1992, 26:105–110.

7. Chaudron L, Szilagyl PG, Kitzman HJ, et al.: Detection of Postpartum Depressive Symptoms by Screening at Well-Child Visits. Paediatrics 2004, 113:551–558.

8. Boyce P, Stubbs J, Todd A: The Edinburgh Postnatal Depression Scale: Validation for an Australian Sample. Aus NZ J Psychiatry 1993, 27:472–476.

9. Areias M, Kumar R, Barros H, et al.: Comparative Incidence of Depression in Woman and Men, during Pregnancy and after Childbirth. Validation of the Edinburgh Postnatal Depression Scale in Portuguese mothers. Br J Psychiatry 1996, 169:30–35.

10. Wickberg B, Hwang C: The Edinburgh Postnatal Depression validation on a Swedish community sample. Acta Psychiat Scand 1996, 94:181–184.

11. Benvenuti P, Ferrara M, Niccolai V, et al.: The Edinburgh Postnatal Depression Scale: Validation for an Italian Sample. J Affec Disord 1999, 53:137–141.

12. Garcia-Esteve L, Ascaso C, Ojuel J, et al.: Validation of the Edinburgh Postnatal Depression Scale (EPDS) in Spanish. J Affect Disord 2003, 75:71–76.

13. Ghubash R, Abou-Saleh MT, Daradkeh TK: The validity of the Arabic Edinburgh Postnatal Depression Scale. Soc Psychiatry Psychiatr Epidemiol 1997, 32:474–476.

14. Adouard F, Glangeaud-Freudenthal NM, Golse B: Validation of the Edinburgh Postnatal Depression Scale (EPDS) in a Sample of Women with High-Risk Pregnancies in France. Arc Womens Men Health 2005, 8:89–95.

15. Werrett J, Clifford C: Validation of the Punjabi version of the Edinburgh Postnatal Depression Scale (EPDS). Int J Nurs Stud 2006, 43:227–236.

16. O'Hara MW: The Nature of Postpartum Depressive Disorders. In Postpartum Depression and Child Development. Edited by: Murray L, Cooper PJ. New York: The Guilford Press; 1997:3–31.

17. Cox JL, Holden JM, Sagowsky R: Detection of the postnatal depression: development of the 10- item Edimburgh Postnatal Depression Scale. Br J Psychiatry 1987, 150:782–786.

18. Matthey S, Barnett B, Kavanagh DJ, et al.: Validation of the Edinburgh Postnatal Depression Scale for men, and comparison of item endorsement with their partners. J Affect Disord 2001, 64:175–184.

19. Glaze R, Cox JL: Validation of a computerised version of the 10-item (self-rating) Edinburgh Postnatal Depression Scale. J Affect Disord 1991, 22:73–77.

20. Onozawa K, Kumar RC, Adams D, et al.: High EPDS scores in women from ethnic minorities living in London. Arch Women Men Health 2003, 6(2):51–55.

21. Matthey S, Barnett B, Ungerer J, Waters B: Paternal and maternal depressed mood during the transition to parenthood. J Affect Disord 2000, 60:75–85.

22. Manzano J, Palacio Espasa F, Zilkha N: Les scènarios narcissiques de la parentalité. Paris: PUF; 1999.

23. Murray L, Cooper PJ: Postpartum Depression and Child Development. New York: The Guilford Press; 1997.

24. Mezzacappa ES: Breastfeeding and maternal stress response and health. Nutr Rev 2004, 62:261–268.

25. Hiltunen P, Jokelainen J, Ebeling H, et al.: Seasonal variation in postnatal depression. J Affect Disord 2004, 78:111–118.

26. Falceto OG, Giuliani ERJ, Fernandes CLC: Influence of Parental Mental Health on Hearly Termination of Breast-Feeding: a Case-Control Study. J Am Board Fam Med 2004, 17:173–183.

Financial Access to Health Care in Karuzi, Burundi: A Household-Survey Based Performance Evaluation

Sophie Lambert-Evans, Frederique Ponsar, Tony Reid, Catherine Bachy, Michel Van Herp and Mit Philips

ABSTRACT

Background

In 2003, Médecins Sans Frontières, the provincial government, and the provincial health authority began a community project to guarantee financial access to primary health care in Karuzi province, Burundi. The project used a community-based assessment to provide exemption cards for indigent households and a reduced flat fee for consultations for all other households.

Methods

An evaluation was carried out in 2005 to assess the impact of this project. Primary data collection was through a cross-sectional household survey of the *catchment areas of 10 public health centres. A questionnaire was used to determine the accuracy of the community-identification method, households' access to health care, and costs of care. Household socioeconomic status was determined by reported expenditures and access to land.*

Results

Financial access to care at the nearest health centre was ensured for 70% of the population. Of the remaining 30%, half experienced financial barriers to access and the other half chose alternative sites of care. The community-based assessment increased the number of people of the population who qualified for fee exemptions to 8.6% but many people who met the indigent criteria did not receive a card. Eighty-eight percent of the population lived under the poverty threshold. Referring to the last sickness episode, 87% of households reported having no money available and 25% risked further impoverishment because of healthcare costs even with the financial support system in place.

Conclusion

The flat fee policy was found to reduce cost barriers for some households but, given the generalized poverty in the area, the fee still posed a significant financial burden. This report showed the limits of a programme of fee exemption for indigent households and a flat fee for others in a context of widespread poverty.

Background

Although the political situation in Burundi has now stabilised, the civil conflict of 1993-2003 deeply affected the country's inhabitants and infrastructure, particularly the health care system. In order to help rebuild the system, a countrywide cost recovery was implemented in 2002 along with a national policy where the communal authorities were supposed to issue exemption certificates for the poorest who could not afford health care costs [1]. Despite this plan, significant financial barriers to access health care continued to exist. A Save the Children report indicated that "There are serious concerns about the effectiveness of this scheme in protecting the most poor from the cost of illness and in improving their access to health services" [2]. A Médecins Sans Frontières (MSF) survey found that almost a million persons were excluded from health care mainly because of financial reasons and only 0.6% had an exemption certificate [3]. Despite this evidence, the health authorities continued to implement the national cost-recovery policy like many low-income countries which still rely on user fees [4].

Given this situation MSF offered to support health authorities in Karuzi province to develop a project to reduce financial barriers to seeking care in government health facilities. MSF's initial proposal of free care was not accepted by the authorities. Consequently, the project combined reduced fees for everyone and a community-based identification exemption system for indigents. The literature shows evidence that fees impact negatively on care utilisation [5,6] and risks pushing people into poverty. James sees fees as a regressive health financing mechanism [7]. The experience of abolishing user fees on utilisation of health services in South Africa or in Uganda showed positive impact [7-9].

Most documented experiences of maintaining user fees whilst trying to mitigate their impact on the poor through individual-targeting waiver schemes performed weakly [10]. However isolated experiences of subsidies to the poor reached up to 40% of the population through health equity funds in Asia [10] and there are examples in the literature of successful experiences in Latin America [11]. Some authors therefore concluded that the weak performance of many waiver schemes stems from poor policy design and underfunding of the projects [10,12].

In our project, MSF subsidised the cost of care through a combination of reduced flat fee for everyone and free care for indigents identified at community level. The exemption system for indigents incorporated previous experience and recommendations, both in terms of design and funding, in an attempt to overcome weak performance of such schemes in other contexts. The exemption system therefore incorporated previous experience and recommendations [12]: correct and timely financing to compensate structures for lost revenue, proper information to the beneficiaries, pre-identification (or active identification) of households in the community, as well as having an international NGO (MSF) playing the role of driving agent and financier.

Objective

This article describes the impact on financial access to health care in Karuzi province, Burundi, of a program using (a) a reduced flat fee for any medical consultation and (b) a community-based assessment to exempt the poorest people from any payment, two years after implementation of the system.

Methods

Setting

Located in Northeast Burundi, Karuzi was a hilly province [13] with 329,431 inhabitants in 2005 [14]. The population was almost all living in a subsistence economy. There were one referral hospital and 13 health centres (HCs), of which three were private, and 10 were government supported, scattered across the province's seven communes.

MSF Project

MSF had been active in this province since 1993. In 2003, MSF began the community project in cooperation with the provincial government and provincial health authority (BPS) to guarantee financial access to quality health care for Karuzi's population through the support of 10 government HCs and the hospital. MSF involvement included provision of drugs and medical materials, reinforcement of HC management and community involvement, supervision, training and monthly financial incentives for the staff and financial support, as described below. These measures were meant to improve both financial accessibility and quality of care to the population.

There were two financial features of the MSF plan that, when combined, were unique when compared to other similar projects. First, for the large majority of the population, an all-inclusive flat fee of 300 BIF (0.28 US$ in August 2005) per visit was applied for whatever care or treatment was received in all public facilities. The amount of 300 BIF was chosen in collaboration with the BPS based on the income of a daily worker in the fields. By January 2003, the flat fee was implemented for the majority of the population, besides indigents, in all the supported facilities. Fees collected by HCs were used to cover part of their operating costs. MSF compensated HCs for lost fee revenue due to this program.

In the same year, communities across Karuzi elected voluntary members of Health Committees (Hcoms) that were broadly representative of the province's demographics (age, gender, socio-economic status, ethnic group). Hcoms were in charge of disseminating information about the program to the community, ensuring correct implementation of the tariff by the health staff and monitoring problems during the project.

The second financial feature was that the poorest households in Karuzi province were identified by a community level assessment to receive an exemption card entitling them to receive free health care in the supported HCs. This exemption system took into account previous experiences and recommendations in different health contexts, such as making sure there was adequate financial compensation for lost (fee) revenues for the HCs, proper information given to the public about the program, and active identification of indigent households in the community [12].

The exemption system was supposed to ensure that identified households (cardholders) would have access to free health care. However, there were concerns that local health authorities, acting alone, would select a limited number of households because of conflicting interests. This conflict in health service objectives between equity and resource generation has been documented in the literature [15,16]. MSF, therefore, wanted an independent group close to the community to decide on exemptions. Since there was no such existing group, as has been described in other contexts like Cambodia [17], MSF asked the Hcoms to make the assessment, as their members had been elected by and were known by the population of the hills. This targeting mechanism has been advocated in the literature on the basis that more and better quality information is available to communities about their members' resources, needs and circumstances [16].

Indigent Household Identification

The Hcoms were charged with identifying indigent households in the communities through socio-economic criteria. These criteria were developed by other Non Governmental Organizations working in the province, together with the provincial authorities (Appendix 1) on the basis of previous assessments [18]. Households meeting at least one indigence criterion were included as beneficiaries. In almost all households, cash availability was low or non-existent so that household income could not be used as an economic criterion as in Cambodia [17] or Thailand [19]. We therefore used the proxy indicators of socio-economic status and vulnerability, such as ownership of land and age and sex of head of household, to capture different aspects of deprivation [16].

Hcom members received training in the community identification procedures by MSF and BPS members. Once they had identified households meeting indigent criteria, random verification by the BPS and MSF was used to avoid nepotism; 10% of all identified households were visited to check their status. The communities also participated in the verification process: MSF and BPS introduced the lists of identified households to meetings with the population where the people could confirm or reject a household's indigent status. Because so many people were poor, stigma attached to receiving a card was low. On the contrary, people hoped to get the cards and avoid financial stress.

Two full-time MSF employees were responsible for listing the identified households, filling out the exemption cards by hand, and organising the administrative system. Exemption cards included information on name, age, sex and "address" of all household members. Hcoms distributed these exemption cards to the identified households. Lists were to be officially updated by Hcoms every two months to add or remove households who had recently become indigent or had improved their situation and no longer qualified. In the end 15% of households (11,247 households) were identified as indigent in early 2005.

Design

The study included the following MSF data sources: a cross-sectional household survey of the population living in the catchment's area of the 10 public HCs (within 5 km) to determine the financial access to heath care (A), and MSF reports from January 2003 to September 2005 (B).

A. Household Survey

Karuzi province had 215,470 inhabitants living in the catchment's area of the 10 HCs supported by MSF [13]. Provinces in Burundi were administratively divided into communes, then further divided into hills and under-hills. People did not live in villages but were scattered amongst the hills and the distance from one house to another could sometimes mean 30 minutes walking.

A three-stage, cluster-sampling survey was carried out between 22 August, 2005 and 14 September, 2005. Population data used for the sample was based on the official registration for each commune in the province in 2000. In the first stage of sampling, population figures were used to determine the location of the clusters in the hills. Surveyed hills were randomly selected, taking into account the relative proportion of their population. We used a random number generator to select hills from administration lists. Because some hills were geographically large, "under-hills" in the selected hills were also selected randomly (second stage of sampling).

In the third stage of sampling, within each of the chosen under-hills, the first household to be surveyed was determined by spinning a bottle in the middle of the under-hill to choose a direction. Interviewers then walked until the limit of the under-hill and counted the number of houses in that direction. A table of random numbers was used to choose the first house to visit. After obtaining consent, the head of that household was interviewed. Subsequent questionnaires were successively administered to the households located directly to the right of the front door of the preceding house until the outer edge of the cluster was reached. In case a household was absent or in case the head of household was not available, teams were instructed to revisit the house later in the day. In case they were still absent or unavailable, the household was replaced by the nearest household.

A household was defined as a group of people who shared the same food and slept under the same roof for at least three nights per week.

Measurement

The questionnaire included 27 items on the following subjects:

- Household composition.
- Household socio-economic indicators: expenses, access to and use of land.
- Possession of an indigent card and indigence criteria met by the household.
- Health-seeking behaviour for the last episode of illness in the household: whether or not a consultation had been sought, place of consultation or reason for non-consultation and place and completeness of medication received.
- Costs of care and source of the money spent for health care costs.

This study was carried out at the same time as a mortality survey whose recall period, based on previous experience and literature, was three months [20]. For the question related to the episodes of illness, the same recall period was chosen-covering the time from the date of the communal election (3 June, 2005—an important reference in the recent history of the country) up to the date the questionnaire was administered (mean recall period = 90.7 days). Households were asked to relate the most recent episode of illness that occurred during this period. There were no unusual events or changes in the environment such as epidemics or problems at the HCs such as major drugs out of stock or strikes by the health staff during this three-month period.

For this study, we defined having "financial access to primary health care" when a person attended the nearest health centre for a consultation and received a full course of medication on site. Note that in this context virtually everyone was prescribed some medication if they attended a HC. Medication given was then described as full or partial based on the declaration of the respondent. In this study, utilisation of health services at a HC near the house of the respondents was used as a proxy indicator for access. In this context, availability of health structures was not an issue as the sampled households lived around HCs supported by MSF. The choice of the sample of people living at a maximum distance of five km from the HC was made to be sure we would get information about other factors influencing health seeking behaviour besides distance. It is clear however, that for the households living further away, distance would be an additional obstacle.

For people who did not attend the nearest health centre, reasons for their alternative choices or for staying at home were recorded. We categorized barriers to use of health services as financial and non-financial. Households were considered as excluded from health care when they did not seek any care although they felt it was necessary.

When there were no episodes of sickness in the household during the recall period, only questions related to household composition and socio-economic situation were asked.

We used expenses as an indicator of socioeconomic status of the households in the analysis. They were a more stable estimate than income in this type of rural economy, where most people were engaged in subsistence farming, and incomes were influenced by season, and were often irregular or under-estimated [21]. Expenses were measured on the basis of consumption in the household the week before the interview.

Additional information about the households' socioeconomic situation was obtained by using "access to a piece of land" to grow crops as an indicator. The households were divided into three groups: people who did not have access to land, people who owned or rented land for subsistence farming, and people who owned or rented land for farming for profit.

Origin of the money covering the cost of care fell into two main categories: households that had money available (savings, business income) for the costs of care at the time of illness and households that had to mobilise the money. For the latter, we considered that a household was:

- in a "precarious situation" if, in order to pay the cost of care, it had to sell a part of its harvest, which normally fed its members, or if someone in the household had to do extra work.
- "impoverished" if the household had to borrow money or if land, animals or part of a future harvest were sold to pay for the consultation.

The questionnaire was translated into Kirundi, back translated into French and pilot-tested with 30 households in July 2005 by the coordinator of the survey

Human Resources

Considering the difficulty of carrying out a survey in such a place, eighteen interviewers and three supervisors were recruited. They were selected on the basis of their education level, their knowledge of the province, their ability to speak French and Kirundi fluently and their physical condition. The team received three days of training on survey methods and procedures. Their understanding of the procedures was evaluated through a half-day test of the questionnaire in hills not selected in the sample. The interviewers were regularly monitored by three supervisors headed by a general coordinator during the survey.

Sample Size

The sample size was calculated using an estimate that 75% of households would have access to primary health care. If the margin of error was fixed at +/- 4% with a cluster effect estimated at 2, then 900 households with at least one sick person in the recall period were required. The calculation used the standard formula: $(1.962 \times 2 \times (0.75 \times 0.25/(0.04)2) = 900$ [22]. Thirty clusters of 30 households with at least one sick person were visited. However, if a household had no sick person in the recall period, another household was added in order to reach our desired sample size.

Data Handling

The data was entered daily by two interviewers/encoders into Epi Info 6.04 and checked by the general coordinator and, if necessary, with interviewers and supervisors. The epidemiological/statistical analysis was carried out in Brussels. Geometric means were calculated for costs and expenses. All confidence intervals were calculated at 95% (95%CI). Proportions were compared by the standard chi-squared tests.

Ethics

Permissions to carry out the studies were obtained by the BPS and the Provincial authority of Karuzi and approved by the Ethics Review Board of MSF.

B. MSF Program Reports

A document review of MSF materials included the monthly reports of the project, monthly HC data and reports of the expatriates working on the health and management committees from January 2003 to September 2005. Those documents complemented the survey data and were useful for describing the setting, the evolution of the project and the resources needed for the exemption system.

We used the documents as supporting qualitative information about context and programme evolution. We did not undertake a formal qualitative analysis.

Results

Description of the Sample

A total of 1031 households were visited, 94 households were absent and 937 households were interviewed, representing 4,949 persons. No household refused to be interviewed. The average size of the household was 5.3 persons (95%CI:

5.1-5.5). The age distribution of the sample was similar to national figure for Burundi [23].

Expenses per person were, on average, 375.0 BIF (335.0-420.7) per week.

There were 86.9% (83.8-89.9) of the households who owned a piece of land and 6.4% (3.9-8.9) rented it to provide subsistence farming for their members. Another 1.3% (0.3-2.3) owned/rented a land for profit and 5.4% (3.3-7.6) of households had no access to land at all.

Results of the Exemption System

Characteristics of Cardholders

Eighty-one households—8.6% (6.3-11.0)—had an exemption card.

Cardholders had lower weekly expenses and fewer owned a piece of land than non-cardholders (Table 1).

Table 1. Comparison between non-cardholders and cardholders

	Households without cards (n = 853*)			Households with cards (n = 81)		
Expenses/pers./week	Amount (BIF) 397.2		CI 95% 352.4-446.7	Amount (BIF) 208		CI 95% 133.7-324.3
	n	%	CI 95%	n	%	CI 95%
Own/rent land for profit	11	1.3%	0.3-2.2	1	1.2%	0.0-3.6
Own land for survival	760	88.90%	85.8-92.0	53	65.40%	53.4-77.5
Rent land for survival	56	6.60%	3.9-9.2	4	4.90%	0.5-9.4
Without land	28	3.30%	1.3-5.3	23	28.40%	16.7-40.1

* 1 missing

Accuracy of the Community Identification Method

Overall, 9.2% (6.1-12.3) of households (86) were found to have indigence criteria as defined in Appendix 1. Thirty-three of them (38%) were landless. As table 2 shows, 44.2% of the households having at least one indigence criterion were cardholders. However, more than half of those with indigence criteria did not have a card (coverage) and almost half of those having a card did not fulfil the criteria (leakage).

Table 2. Distribution of cardholders/positive indigence criteria

	Households with card			Households without card		
	n	%	CI 95%	n	%	CI 95%
With indigence criteria	38	44.20%	26.4-62.0	48	55.80%	38.0-73.6
Without indigence criteria	43	5.10%	3.7-6.4	808	95.00%	93.6-96.3

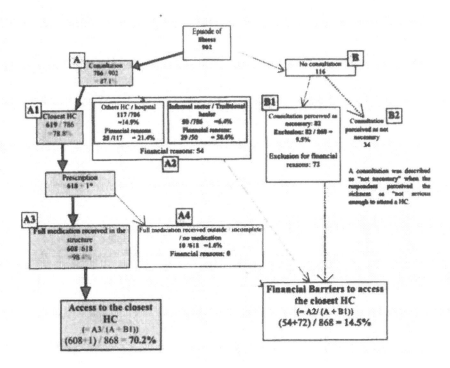

* No prescription necessary as diagnosis not clear at HC

Figure 1. Access to primary health care.

Access to Health Care

In the 937 households surveyed, 902 cases of illness were reported (96.3%). As seen in Figure 1, 87.1% (84.5-89.8) went for a consultation in various locations:

- 78.8% (73.4-84.1) sought medical help at the closest HC (supported by MSF applying the flat fee/cards exemption).

- 12.1% (6.7-17.5) went to another HC and 2.8% (0.5-5.1) went directly to the hospital. Of these, 21.4% (2.9-39.8) explained their choice was due to financial barriers. Other barriers included a lack of confidence with the HC staff or a shortage of medication at the HC.

- 6.4% (4.5-8.2) went to the informal sector/traditional healers, 58.0% (45.2-70.8) claimed to do so for financial reasons.

Of those who consulted in the closest HC, 98.4% (97.3-99.4) received full treatment.

In total, 70.2% (64.9-75.5) of the population had access to care at their local HC.

However, 9.5% (7.0-12.0) of the households were excluded from health care and for 87.8% of them (76.1-94.5); this was due to financial barriers. Overall, 14.5% (10.5-18.5) of households experienced financial barriers that prevented them from accessing care at their closest HC.

Access to Health Care for Cardholders

Of the cardholders, 91.1% (72/79) went for a consultation for the last episode of sickness and 88.9% of these (64/72) went to the closest HC. Of those going to the closest HC, 96.9% (62/64) obtained full treatment but four of the cardholders paid 300 BIF. Access to health care among cardholders was higher than non-cardholders; 80.5% (CI 95%: 72.4-88.6) vs. 69.2% (CI 95%: 63.3-75.0). Although this difference did not reach statistical significance, the relative risk was significant; cardholders had 1.58 (0.99-2.54) more chance to access health care than non-cardholders.

Costs of Care

Of the people who went for a consultation at the closest HC (n = 619),

- 12.0% (8.8-15.1) did not pay for health care (74 cases) and 81.1% (71.8-98.3) had a card (60 cases).

For those who paid,

- The average amount paid was 311.2 BIF (304.8-317.7) or median amount of 300 BIF.
- 6.3% (1.8-15.7) of the cardholders (4 cases) still paid the flat fee.
- 2.5% (1.2-3.8) of non-cardholders (14 cases) did not pay for health care.

Origin of the Money and Effects of the Cost of Care on Households

Of the cases who went for a consultation at the closest HC, Table 3 shows that only 13.1% of the households had the money available at the moment of the disease.

Of the rest, 61.0% entered into a precarious situation because they had to look for extra work outside of the household (123 cases) or they had to sell a part of their harvest that they would normally have used for their own food (209 cases) and 25.9% of the households became impoverished. Most of them went into debt

(116 cases), some had to sell part of their future harvest (20 cases) and a few sold animals or land (5 cases).

Table 3. Situation of households that paid-for health care

	n = 544*	%	CI 95%
Money available	71	13.10%	9.0-17.1
Precarious	332	61.00%	54.3-67.8
Impoverished	141	25.90%	20.3-31.6

* 1 missing

Discussion

This study showed that despite an innovative program that combined a flat fee for care for most people with an exemption system for indigents, there continued to be significant financial barriers in accessing health care in Karuzi. Even though access to health care was readily available for 70% of overall population, about 30% sought care elsewhere or stayed home. Why was this so?

Significant Overall Poverty

A key feature was that the national poverty line in rural areas was evaluated at 1031.73 BIF/person/week [24], so that 88.0% (83.5-91.6) of the population in Karuzi lived below that threshold at the time of the survey, confirming a context of widespread generalized poverty. Cash available at the household level was limited and it was difficult to mobilise quickly in the face of unexpected expenses for an episode of illness.

Although most households in the province owned a piece of land, it was just enough for subsistence, not for profit. Assessments carried out at the national level revealed that the average size of land per household was less than 0.5 ha [25], barely enough to feed household members, let alone generate income. The situation became worse during the crisis that affected the country during the past decade when many households had to sell their land [25]. "Landless" was the main indigence criterion reported in the survey.

The Exemption System: Mixed Results

The exemption system was found to protect only 8.6% (6.3-11.0) of the population. This was lower than the 15% of households identified by the Hcoms but larger than in other provinces in Burundi [3] or in other countries where classical exemption mechanisms were in place [26-28]. Exemption schemes, even those endorsed by policy and legislation, are rarely fully effective. Schemes aimed at targeting the poor with exemptions often miss the intended beneficiaries, those in greatest need [29].

MSF studies have shown that classical exemption mechanisms in Mali, Sierra Leone or Haiti covered less than 2% of the population" [30]. And Ridde [31] noticed that, although exemption of payment for indigents was one of the core principles of the Bamako Initiative, in many settings, exemptions had not been implemented or had not been able to protect very many people in the community.

However, results similar to those of our study were found with equity funds in Cambodia [29,32]. Their design was close to the exemption system developed in Karuzi and included the existence of donor funding, the presence of a driving agent, clear separation of roles, and appropriate identification techniques. These studies reached conclusions similar to ours: while these mechanisms are superior to traditional waiver systems in terms of health services utilisation by the target group, studies reveal remaining barriers to access and indebtedness prevention.

However, existing studies tend to focus on hospital and health structure data while failing to provide information on non-users of the system. Our study therefore brings additional data on general population access. These data are crucial to be considered for national policy choices.

Our results showed that cardholders had better financial access to health care than non-cardholders. It was also encouraging that almost all cardholders benefited from free health care in public HCs.

Although most of the cardholders were in a worse socio-economic situation than the rest of the population, the exemption system had problems properly covering the poor as shown by 56% of the households who had at least one indigent criterion but did not get cards. One reason for this low coverage could be the time lag between the date of identification and the survey. This would reinforce the idea that exemption systems do not capture properly the dynamic dimension of poverty. At the same time, 5% of households did get cards despite not having met the criteria. As well, although Hcoms had identified 15% of households as indigent, our survey found only 8.6% of households in possession of a card. This difference could be explained by the following:

- Around 400 cards (3.5% of all the cards distributed) were never distributed to the identified households because the householders were absent during the distribution meetings and the Hcoms did not follow up.

- Initially, some health staff members were reluctant to accept the exemption system and confiscated cards when they considered that beneficiaries did not qualify as indigent. This situation improved with time thanks to the reporting of the Hcoms.
- Cards lost were often not renewed by the Hcoms.

These difficulties should have been handled by the Hcoms with updates to the list every two months. However, this would have implied constant and time-consuming re-assessment. Experience in this project showed that maintaining an accurate list was a complicated and imperfect process, very demanding both financially and in terms of human resources. Members of Hcoms were not paid and this may have affected their motivation.

Furthermore, the performance of the system was hampered by the difficulty in defining and interpreting indigence criteria in a context of generalised poverty. The feasibility and accuracy of distinguishing the poor from the non-poor to determine eligibility for exemptions, is fraught with problems [28,33]. The notion of indigence is complex and covers both poverty and social exclusion [34]. Criteria might have been subjectively and arbitrarily interpreted even though the members of the health committees had been trained to recognise eligible households. "A major difficulty is to identify very poor people in a population in which the poverty is rife" [33].

The exemption system was demanding in terms of other human and financial resources. In addition to the Hcom members' time, human resource inputs included significant time of two full-time staff members, financed by MSF. Significant financial costs arose from MSF's financial support to HCs for compensation of revenue lost due to exemptions, considered a key condition for fee exemption schemes to be pro-poor [12]. This raises the question of whether such resources should be provided to a system which benefits a relatively small number of people, particularly in a context of poverty. Other authors argued that "A universal free healthcare approach is justified in all situations with widespread misery or when time does not allow individual assessment schemes to be implemented [...]. Alternatively, identifying people living in poverty (by proxy means testing) and targeting benefits to them could be more attractive than a universal approach if the proportion of poor people in the society is not overwhelming..." [10]

Access to Health Care

Despite these difficulties there were some positive elements to the project. In 2003, for all of Burundi, the level of access to health care in areas where cost-recovery was implemented was 58% of the population [3]. In the Karuzi project, access to health care was much better, as 70% of the population went for a consultation to the closest HC and received full medication. Other encouraging results were that consultations in the informal sector were few (6.4%) as compared to other African contexts [30].

Although mechanisms set up in the project to improve access proved to have done so compared to other areas in Burundi where cost recovery schemes were implemented, they still revealed important limitations in their potential to increase the use of health services for the population. Almost 15% still had financial barriers to access the closest HC and 10% of the population remained excluded from health care mainly for financial reasons. These findings were in line with other surveys revealing that even low fees constituted an obstacle to patients' access in contexts of widespread poverty [30,35]. MSF's experience has also revealed that targeting strategies compared poorly to general exemption or those based on large categories (like women and children), such as those implemented in Burundi in 2006 (national free care policy for under-5s and pregnant women). In addition, in contexts where health fees were totally abolished for all patients, evidence has shown an increase in the use of health services and specific benefits for the poorest households [35].

Cost of Care, Financial Burden in Households

A specific feature of the project was the implementation of the reduced flat fee system. Although the amount of the agreed fee seemed minimal and was respected by the staff, one visit to the primary health care level represented almost the equivalent of one day of household expenses. Eighty-seven percent of households reported not having sufficient money available to seek immediate care for their last illness episode. To finance health care, more than a quarter became impoverished and 60% needed extra work to provide cash flow, which risked delaying consultation. Thus, despite the reduced flat fee, the cost of health care still represented an important burden for many households. Further, our study did not include indirect costs linked to treatment such as transportation, food expenditure and loss of time [35]. These costs can initiate a vicious circle underlined by Noirhomme et al. [32] in which "poverty not only brings ill-health, but ill-health also tends to worsen poverty."

Limitations

There were a number of limitations to the study. MSF was well known in the province especially for the identification of indigents. Although the surveyors—MSF employees—clarified that they were not in charge of identifying indigents in the community, respondents might have answered questions in a way to maximise their benefit, for instance, by overestimating their expenses in hope of being included on the indigent lists. This factor is likely to have been limited given the very low level of expenses reported. If present, this bias would lead to an underestimation of real poverty levels rather than an overestimation, as expenses were used to assess the socio economic status of households. As well, confronted by a western medical organisation, the respondents might have underreported the use of traditional medicines or of the informal sector.

An additional concern may be the 90.7-day recall period for last illness episode, which may have reduced the accuracy of details recalled. However, potential recall bias was reduced since the illness of inquiry was the most recent during the recall period.

Conclusion

An innovative approach of adopting a flat fee for consultations in primary care clinics in rural Burundi and identifying indigents was somewhat successful in increasing financial access to care. However, against a background of widespread poverty, and the difficulties to properly target the poor in such a context, many people still did not obtain appropriate care or suffered financial hardship doing so. Introducing these measures was cumbersome, and not really responsive to the poverty dynamics in the population. These results indicate that alternative strategies, such as free care for everyone or targeted groups such as under-5s or pregnant women as implemented in Burundi in 2006, are needed to ensure increased access to effective health services for the poor.

Competing Interests

The authors declare that they have no competing interests.

Authors' Contributions

SLE, FP, CB, MVH, TR, and MP contributed to study conception and design, interpretation of the data, and drafting and revising the manuscript; SLE and FP led data collection; CB led the statistical analysis with additional analysis by SLE and MVH. All authors read and approved the final manuscript.

Appendix 1 – Exemption Criteria

Groups Identified by Hcom

° Long-term welfare recipient: Socially isolated, without children, often elderly without family. No income or savings; inadequate food, clothes and accommodation. Lives with the support of the neighbourhood. Often disabled and/or with serious health problems.

° Without land/property: live in hut, seasonal workers.

° Disabled people in household with no member capable of working, not receiving assistance from outside the household.

° Elderly (over 55 years old) in household with no member capable of working, not receiving assistance from outside the household.

° Orphans head of household (below 18 years old) not receiving assistance from outside the household.

° Foster care and orphans care, taking care of indigents, in household with no member capable of working, not receiving assistance from outside the household.

° Widowed, head of household, in a household with no member capable of working; low income; not receiving assistance from outside the household.

Groups Identified by HCR, Directors of School

° Repatriated households, 6 months after their return.

° Students meeting above criteria.

Acknowledgements

The authors thank the provincial government and provincial health authority of Karuzi, Burundi for their partnership and the survey teams, as well as the population of Karuzi province, for their participation in this project and its evaluation.

References

1. République du Burundi. Ministère de la Santé publique et Ministère de l'intérieur et de la sécurité publique: Ordonnance ministérielle n° 630630/445

du 02/04/2003 portant fixation des modalités de prise en charge médico-sanitaire des indigents. Bujumbura. 2003.

2. Save the Children UK: An Unnecessary Evil? User Fees for Healthcare in low-income Countries. The Cost of Coping with Illness. Burundi. Briefing. London. 2003.

3. MSF-Belgium: Access to healthcare in Burundi: results of three epidemiological surveys. Research and analysis report. Brussels. 2004.

4. World Health Organisation: The World Health report 2008. Primary Healthcare. Now ore than Ever. Geneva. 2008.

5. Creese AL: User charges for health care: a review of recent experience. Health Policy Plan 1991, 6(4):309–19.

6. Palmer N, Mueller DH, Gilson L, Mills A, Haines A: Health financing to promote access in low income settings - how much do we know? Lancet 2004, 364(9442):1365–70.

7. James C, Hanson K, Mcpake B, Balabanova D, Gwatkin D, Hopwood I, Kirunga C, Knippenberg R, Meessen B, Morris SS, Preker A, Souteyrand Y, Tibouti A, Villeneuve P, Xu K: To retain or to remove user fees? Reflections in the current debate in low and middle income countries. Applied Health Econ Health Policy 2006, 5(3):137–153.

8. Wilkinson D, Gouws E, Sach M, Karim SS: Effect of removing user fees on attendance for curative and preventive primary health care services in rural south Africa. Bull World Health Organ 2001, 79(7):665–71.

9. Nabyonga J, Desmet M, Karagami H, Kadama PY, Omaswa FG, Walker O: Abolition of cost sharing is pro-poor: evidence from Uganda. Health Policy Plan 2005, 20(2):100–8.

10. Sepheri A, Chermonas R: Are user charges efficiency and equity-enhancing? A critical review of economic literature with particular reference to experience from developing countries. J Int Dev 2001, (13):183–209.

11. Gwatkin DR: The current state of knowledge about Targeting health programmes to reach the poor. World Bank. Washington DC; 2000.

12. Bitràn R, Giedion U: Waivers and Exemptions for Health Services in Developing Country. In Social Protection Discussion Paper Series. Washington DC: The World Bank; 2003.

13. Protopopoff N, Van Herp M, Maes P, Reid T, Baza D, D'Alessandro U, Van Bortel W, Coosemans M: Vector control in a malaria epidemic occurring within a complex emergency situation in Burundi: a case study. Malar J 2007, 16(6):93.

14. Protopopoff N, Van Bortel W, Marcotty T, Van Herp M, Maes P, Baza D, D'Alessandro U, Coosemans M: Spatial targeted vector control in the highlands of Burundi and its impact on malaria transmission. Malar J 2007, 3(6):158.

15. Hardeman W, Van Damme W, Van Pelt M, Por I, Kimvan H, Meessen B: Access to health care for all? User fees plus a Health Equity Fund in Sotnikum, Cambodia. Health policy Plan 2004, 19(1):22–32.

16. Hanson K, Worrall E, Wiseman V: Targeting services towards the poor: a review of targeting mechanisms and their effectiveness. In Health, economic development and household poverty: from understanding to action. Edited by: Mills A, Bennett S, Gilson. Published by Systems Resource Guide. London: Routledge; 2006.

17. Bart J, Neil P: Improving access for the poorest to public sector health services: insights from Kirivong Operational Health District in Cambodia. Health Policy and Plan 2006, 21(1):27–39.

18. Africare/Burundi: Mise en oeuvre de la composante "développement communautaire" du PDRDMR en province de Karusi. Rapport sur le ciblage de la pauvreté dans la province de Karuzi. Washington, Bujumbura. 2001.

19. Mills A: Exempting the poor: the experience of Thailand. Soc Sci Med 1991, 33(11):1241–52.

20. Checchi F, Robert L: Interpreting and using mortality data in humanitarian emergencies. Volume 52. Humanitarian practice network, ODI; 2005.

21. Dercon S: Poverty measurement. In The Elgar Companion to Development Studies. Edited by: Clark DA. Cheltenham: Edward Elgar; 2006.

22. Kish L: Survey Sampling. New York: John Wiley & Sons, Inc; 1965.

23. République du Burundi. Ministère de la Santé Publique: Plan National de Développement Sanitaire 2006–2010. Bujumbura. 2005.

24. République du Burundi: Poverty Reduction Strategy Paper. Bujumbura. 2003.

25. Centre d'Alerte et de Prévention des Conflits (CENAP): Pratiques rurales en matière de gestion des propriétés foncières, rapport final. Bujumbura. 2005.

26. McPake B, Hanson K, Mills A: Experience to Date of Implementing the Bamako Initiative: a review and five Country Case Studies. London School of Hygiene and Tropical Medicine. London; 1992.

27. Russell S, Gilson L: User fee policies to promote health services access for the poor: a wolf in sheep's clothing? Int J Health Serv 1997, 27(2):359–79.

28. Gilson L: The lessons of user fee experience in Africa. Health Policy Plan 1997, 12(4):273–85.

29. Jacobs B, Price NL, Oeun S: Do exemptions from user fees mean free access to health services? A case study from a rural Cambodian hospital. Trop MedI Int Health 2007, 12(11):1391–401.

30. MSF-Belgium: No cash, no care: how "user fees" endanger health. An MSF briefing paper on financial barriers to Healthcare. Brussels. 2008.

31. Ridde V: L'initiative de Bamako 15 Ans Après. Un Agenda Inachevé. In Health, Nutrition and Population Discussion Paper. Edited by: Preker AS. Washington DC: The World Bank; 2004.

32. Noirhomme M, Meesen B, Griffiths F, Ir P, Jacobs B, Thor R, Criel B, Van Damme W: Improving access to hospital care for the poor: comparative analysis of four health equity funds in Cambodia. Health Policy Plan 2007, 22(4):246–262.

33. Whitehead M, Dahlgren G, Evans T: Equity and health sectors reforms: can low income countries escape the medical poverty trap? Lancet 2001, 358(9284):833–836.

34. Kaddar M, Schmidt-Ehry B, Stierl F, Tchicaya A: Indigence et Accès aux Soins de Santé en Afrique Sub-Saharienne: Situation et Perspectives d'action. Eschbron: GTZ 1997.

35. Meessen B, Van Damme W, Kirunga Tashobya C, Tibouti A: Poverty and user fees for public health care in low-income countries: lessons from Uganda and Cambodia. Lancet 2006, 368(9554):2253–57.

Intention to Breastfeed and Awareness of Health Recommendations: Findings from First-Time Mothers in Southwest Sydney, Australia

Li Ming Wen, Louise A. Baur, Chris Rissel,
Garth Alperstein and Judy M. Simpson

ABSTRACT

Background

In 2001, the World Health Organisation (WHO) recommended exclusive breastfeeding for the first six months of life. The objectives of this study are to assess awareness of the WHO recommendation among first-time mothers (women at 24 to 34 weeks of pregnancy) and to explore the relationship between this awareness and mothers' intention to exclusively breastfeed for six months.

Methods

This study was part of the Healthy Beginnings Trial (HBT) conducted in southwest Sydney, Australia. We analysed cross-sectional baseline data of the trial conducted in 2008, including 409 first-time mothers at 24 to 34 weeks of pregnancy. The mothers' awareness of the recommended duration of exclusive breastfeeding and their intention to meet the recommendation were assessed through face-to-face interviews. Socio-demographic data were also collected. Factors associated with awareness of the recommendation, or the intention to meet the recommendation, were determined by logistic regression modeling. Log-binomial regression was used to calculate adjusted risk ratios (ARR).

Results

Sixty-one per cent of mothers knew the WHO recommendation of exclusive breastfeeding for six months. Only 42% of all mothers intended to meet the recommendation (breastfeed exclusively for six months). Among the mothers who knew the recommendation, 61% intended to meet the recommendation, compared to only 11% among those mothers who were not aware of the recommendation.

The only factor associated with awareness of the recommendation was mother's level of education. Mothers who had a tertiary education were 1.5 times more likely to be aware of the recommendation than those who had school certificate or less (ARR adjusted for age 1.45, 95% CI 1.08, 1.94, $p = 0.02$). Mothers who were aware of the recommendation were 5.6 times more likely to intend to breastfeed exclusively to six months (ARR adjusted for employment status 5.61, 95% CI 3.53, 8.90, $p < 0.001$).

Conclusion

Awareness of the recommendation to breastfeed exclusively for six months is independently associated with the intention to meet this recommendation. A substantial number of mothers were not aware of the recommendation, particularly among those with low levels of education, which is of concern in relation to promoting breastfeeding. Improving mothers' awareness of the recommendation could lead to increased maternal intention to exclusively breastfeed for six months. However, whether this intention could be transferred into practice remains to be tested.

Trial Registration

HBT is registered with the Australian Clinical Trial Registry (ACTRNO12607000168459)

Background

Breastfeeding has a wide range of health benefits for mothers and children and is a key protective factor against childhood overweight and obesity [1-4]. The current World Health Organization (WHO) recommendation for breastfeeding is that all infants should be exclusively breastfed for the first six months of life, and receive nutritionally adequate and safe complementary foods while breastfeeding continues for up to two years of age or beyond [2]. The WHO recommendations have been adopted and endorsed by many countries including Australia [3,5].

In Australia, a national survey found that in 2004-5, breastfeeding initiation was 88% [6], and similarly in the state of New South Wales (NSW), the percentage of infants "ever breastfed" was 90% in 2001 [7] and 87% in 2003-4 [8] respectively. However, this high initiation rate of breastfeeding does not lead to a high prevalence of sustained breastfeeding: only 16% of infants were exclusively breastfed to six months and 29% were breastfed to 12 months [8]. The Health Department of NSW recommends the promotion of breastfeeding for all mothers and infants to focus on extending the duration of breastfeeding to 12 months and exclusive breastfeeding to six months, in particular among those mothers who are most socio-economically disadvantaged, less than 25 years of age, or with less than a tertiary education [8].

Breastfeeding decisions and practices are influenced by multiple factors including knowledge, attitudes and beliefs, as well as socio-cultural and physiological factors [9-13]. However, results from research into determining these factors to date have been very variable due to a lack of objective, reliable, valid and sensitive measures [9]. In developed countries like Australia, mothers who are younger (under 25 years old), have less education, or are most socio-economically disadvantaged tend to have lower rates of full breastfeeding, rates of initiation and duration of breastfeeding [8,14].

There is increasing recognition of the need to promote exclusive breastfeeding since the WHO recommendation on exclusive breastfeeding for six months was made in 2001. However, the impact of such promotion on rates of exclusive breastfeeding is less clear [15]. This might be explained by a poor understanding of the breastfeeding recommendation, and of knowledge, attitudes and practice about breastfeeding in the community.

Research has repeatedly found that women's pre-birth breastfeeding intentions are a good predictor of the actual duration of breastfeeding [16,17]. In a study conducted in a group of Australian women, Rempel found that a strong desire to breastfeed was positively associated with breastfeeding at six months and having no intention to breastfeed was negatively associated with breastfeeding at six months [18].

To enhance breastfeeding promotion strategies in the context of relatively recently changed recommendations, it is important to have a good understanding of mothers' knowledge of the current recommendation on breastfeeding and their intention to meet the breastfeeding recommendation. In 2008, we commenced the Healthy Beginnings Trial (HBT) to test the effectiveness of an early childhood obesity intervention in the first two years of life [19]. The intervention uses a home-visiting strategy to promote healthy feeding of babies among first-time mothers. As part of this trial, we aimed to increase exclusive breastfeeding for the first six months among participating mothers.

This paper reports on those aspects of the data collected for the HBT that were collected at the baseline interview, prior to randomisation. We aimed, firstly, to assess first-time mothers' awareness of the recommended duration of exclusive breastfeeding and their intention to meet this recommendation, and, secondly, to explore the factors that are associated with the intention to exclusively breastfeed so that appropriate breastfeeding intervention strategies could be developed.

Methods

Study Design

The design of the main study is a Randomised Controlled Trial (RCT), however for the purpose of this analysis we have used the data collected at the study baseline, which could be considered a cross-sectional survey. The RCT was conducted in southwest Sydney, Australia in 2008 and approved by the Ethics Review Committee of Sydney South West Area Health Service (RPAH Zone). The details of the HBT research protocol have been reported elsewhere [19].

Study Participants

All pregnant women who attended antenatal clinics of Liverpool and Campbelltown Hospitals located in south-western Sydney were approached by research nurses with a letter of invitation and information about the study. Women were eligible to participate if they were aged 16 years and over, were expecting their first child, were between weeks 24 and 34 of pregnancy, were able to communicate in English and lived in the local areas. Once eligibility was established and consent obtained, women then were asked to fill in a registration form with their contact information to allow the nurses to make further arrangements for the baseline data collection and random allocation to study group.

From around 2700 mothers who were approached, a total of 667 first-time mothers at 24-34 weeks of pregnancy were recruited for the main study. Four hundred and nine mothers were interviewed at their home before giving birth at the baseline and were included in this particular study. Another 258 mothers who also participated in the HBT were excluded, as we were not able to conduct the survey before they gave birth.

Data Collection and Key Measures

A face-to-face interview with participating mothers was conducted by one of four research nurses at their home, prior to randomisation. The interview lasted 20 to 30 minutes and included a range of questions in relation to the general health, physical activity and nutrition of the mothers, as well as demographic information.

To assess mothers' awareness of the breastfeeding recommendation and their intended duration of exclusive breastfeeding, they were asked the following questions:

"What do you understand to be the recommended age to which you should continue to exclusively breastfeed your child?"
"Do you plan to breastfeed your child?"
"To what age do you plan to exclusively breastfeed your child?"

In addition, the mothers were asked the main reasons for their decision to breastfeed, or not to breastfeed with an open-ended question. The face validity of the questions had been pilot-tested by some mothers and reviewed by breastfeeding experts in the field.

Other study variables, including age, employment status, education level, marital status, language spoken at home, and country of birth, were asked using the standard questions from the NSW Health Survey [20].

Analysis

Statistical analyses were carried out using the computer package Stata 10 [21]. Relationships between study and outcome variables were examined using Pearson chi-square tests and Mantel-Haenszel chi-square tests for trend in proportions. Two logistic regression models were developed, one for awareness of breastfeeding recommendation and one for intention to meet the recommendation. Variables that were significant ($P < 0.05$) on bivariate analysis were entered into each model, then the least significant terms were progressively dropped until only those with $P < 0.05$ and those that confounded the effect of these variables remained in the model. Adjusted risk ratios (ARRs) with 95% confidence intervals were calculated by refitting the final models using log-binomial regression with the Stata binreg command.

Results

The main characteristics of the participating mothers are shown in Table 1. The age range of the mothers was from 16 to 46 years with a mean age of 26 years. Most of the mothers (87%) were either married or living with their de facto partner. Twenty three percent had completed tertiary education and 11% spoke a language other than English at home. In addition, 21% were unemployed and 19% had paid maternity leave. Among those mothers who were in the workforce or studying, 11% planned to return to work or study after giving birth within three months, and a further 21% planned to do so within four to six months after giving birth.

Table 1. Characteristics of the 409 participating women and factors associated on bivariate analysis with awareness of the recommendation of exclusive breastfeeding for six months or the intention to meet the recommendations

Among all 409 participating mothers, 61% knew the recommendation of exclusive breastfeeding for the first six months of life, and 39% either did not know or answered incorrectly. Only 42% of all mothers intended to meet the recommendation, however 94% (384) of the mothers did plan to initiate breastfeeding. Among the mothers who knew the recommendation, 61% intended to meet the recommendation, compared to only 11% among those mothers who were not aware of the recommendation.

Table 1 also shows factors associated on bivariate analysis with awareness of the recommendation of exclusive breastfeeding for six months or the intention to meet the recommendations. Awareness of the recommendation of exclusive breastfeeding for six months was significantly associated with older maternal age (Mantel-Haenszel $\chi^2 1$ = 14.9, p < 0.001), marital status (married or de facto) ($\chi^2 1$ = 7.9, p = 0.005), and a higher level of education (Mantel-Haenszel $\chi^2 1$ = 20.3, p < 0.001). Marital status and employment status of the mothers were also found to be associated with their intention to meet the recommendation of exclusive breastfeeding for 6 months ($\chi^2 1$ = 8.9, p = 0.003 and $\chi^2 6$ = 14.8, p = 0.02 respectively). Awareness of the recommendation was very strongly associated with mothers' intention to breastfeed exclusively for six months ($\chi^2 1$ = 103.6, p < 0.001).

The only factor associated with awareness of the recommendation on multivariate analysis was mother's level of education (Table 2). Mothers who had completed university/tertiary education were more likely to be aware of the breastfeeding recommendation than those who had school certificate or less, with an ARR after adjusting for the confounding effect of age of 1.45 (95% CI 1.08, 1.94, p = 0.02). More importantly, awareness of the recommendation was the only factor that was significantly associated with the intention to exclusively breastfeed for six months. Mothers who were aware of the recommendation were 5.6 times more likely to intend to breastfeed exclusively to six months (after adjusting for the confounding effect of employment status, ARR 5.61, 95% CI 3.53, 8.90, p < 0.001). Marital status was dropped in the final model as it was not significant after adjusting for education (p = 0.32), or after adjusting for awareness of the recommendation of exclusive breastfeeding (p = 0.27).

The main reasons given by the mothers for planning to breastfeed or not breastfeed are summarised into several themes, showing some representative examples, in Table 3. The majority of the mothers understood that breastfeeding is good for the baby's and mother's health. For example:

"Breastfeeding is best, nutritious, convenient, cheaper and (helps the) bonding between mother and child."

"Breastfeeding can help mother back to her normal weight quicker, prevent breast cancer and boost the immune system."

."..Good for baby. God made milk special for a baby, (it's) natural, protects the baby from illness."

Table 2. Factors associated in multivariate analysis with awareness of recommendation of exclusive breastfeeding for six months or intention to meet the recommendation

Variables	Awareness of the recommendation of exclusive breastfeeding			Intention to meet the recommendation		
	ARR[**]	95%CI	P	ARR[**]	95%CI	P
Age			0.24[a]			
25-29	1					
<20	0.73	0.49 - 1.01				
20-24	0.96	0.79 - 1.18				
30-34	1.05	0.86 - 1.29				
≥35	1.16	0.93 - 1.44				
Employment status						0.19[c]
Unpaid maternity leave				1		
Employed				1.24	0.96 - 1.62	
Paid maternity leave				0.98	0.76 - 1.29	
Unemployed				0.72	0.50 - 1.01	
Home duties				1.06	0.80 - 1.40	
Student				0.99	0.58 - 1.71	
Other				0.61	0.25 - 1.51	
Education			0.02[a]			
Completed primary school to school certificate	1					
HSC to TAFE certificate[*]	1.20	0.91 - 1.58				
University or tertiary education	1.45	1.08 - 1.94				
Aware of the recommendation of exclusive breastfeeding[*]						< 0.001[d]
No				1		
Yes				5.61	3.53- 8.90	

[*] HSC = Higher School Certificate, TAFE = Technical And Further Education
[**] ARR = adjusted risk ratio
[a] adjusted for education
[b] adjusted for confounding effect of age
[c] adjusted for awareness of the recommendation of exclusive breastfeeding
[d] adjusted for confounding effect of employment status

Mothers who did not plan to breastfeed (6%) gave reasons for not breastfeeding such as: "No time, have to work"; "I would be embarrassed in public" and "Can't stand the thought of it. Freaks me out."

Among the 25 mothers who did not plan to breastfeed, 21 were less than 24 years old, 10 were unemployed, only 15 completed the school certificate or less and only 9 knew the six months exclusive breastfeeding recommendation.

Table 3. The main reasons given by the women for planning to breastfeed or not breastfeed

Reasons for breastfeeding	Number (%) N = 384	Examples of what women said
Baby's health	163 (42)	Health, strength of immune system. Better for baby, everything right for baby. Healthy and nutritious for baby and less infection. Baby gets all nutrition. Baby's health, better than formula. Safest way to feed baby. More nutritious for baby, healthier. Formula can have chemicals and preservative.
Mother and baby's health	38 (10)	Better for baby, benefits for mother: weight loss, contraction of uterus. Good for mother and child. Prevent disease. Help prevent breast cancer in future. Antenatal class changed my mind. Back to normal weight quicker. Protector for breast cancer, can help prevent it. Boost immune system.
Cost effective	52 (14)	Good for baby. Cost effective. Best for baby, immunity, cheaper. Better for baby with immunity. Does not cost anything.
Convenient	29 (8)	Excellent for baby's health. Convenient, immunity, no sterilizing, walking milk bar! Convenient, always there, no sterilization or preparation. Good for baby.
Bonding	44 (11)	Baby's health, bonding. Brings mother and baby closer, healthiest food for the baby. Good for baby, immunity, relationships with mother, growing, bonding.
Nature	35 (9)	Good for baby. God made milk special for a baby, natural, protect the baby from illness. That's what God planned for us to do. It's nature, all other animals do it. Best for baby. More nutritious. More benefits than formula. Mother generally prefers natural remedies.
All of them	25 (6)	Breastfeeding is best, nutrition, convenience, cheaper, bonding. Health benefits for baby. Convenience. Financial benefit. Natural. Good for baby, reduced risk of breast cancer, bonding, cheaper. Natural, cheaper, beneficial for baby, easier.
Reasons for not breastfeeding	**Number (%) N = 25**	
Back to work/no time	7 (28)	No time, have to work. Need to get back to work. No interest, going back to work soon.
Health concerns	2 (8)	Have an infection, unable to breastfeed. I would like to breastfeed.
Uncomfortable/embarrassment	8 (32)	I don't think I will feel comfortable. Too many people around. Did not feel comfortable. I would be embarrassed in public.
Convenient	3 (12)	It's easier to bottle feed. I don't know much about breastfeeding. My family all been bottle fed. More time consuming, hard enough already. It is easier to heat up a bottle.
Just don't want to breastfeed	5 (20)	Just did not want to breastfeed. Can't stand the thought of it. Freaks me out.

Discussion

In this cross-sectional analysis we used data collected antenatally from 409 first-time mothers participating in the Healthy Beginnings Trial in southwestern Sydney. We found that a substantial proportion of the mothers (39%) did not know the current recommended duration of exclusive breastfeeding, and only 41% intended to meet the recommendation to breastfeed exclusively for six months. Awareness of the breastfeeding recommendation was significantly associated with the mother's intention to exclusively breastfeed her child. However, whether the intention to breastfeed is transferred to practice remains to be tested in future research.

Breastfeeding is widely acknowledged to have health benefits for mothers and babies [1-4]. Our study showed that most mothers were aware of some benefits of breastfeeding for both mothers and babies and, indeed, 94% of the mothers planned to breastfeed their child. In contrast, a small proportion of mothers who did not plan to breastfeed (6%) had strong negative attitudes towards breastfeeding. Changing negative feelings or negative perceptions of breastfeeding in this group of mothers is a challenge for breastfeeding promoters, but an important one, because they face greater health risks than the general community, being younger, less well educated and more likely to be unemployed.

To our knowledge, to date there is no research into mothers' awareness of the WHO breastfeeding recommendation and its association with the antenatal intention to breastfeed. Since the intention to breastfeed is a positive predictor for actual duration of breastfeeding [16-18], exploration of the factors influencing the intention may help health workers to address the issues related to breastfeeding intentions. A study by Forster et al revealed that breastfeeding intention was a strong indicator for breastfeeding initiation and duration across all groups of Australian women, including those with less formal education, younger women and those with less social support [16]. Therefore, focusing on mothers' intention to breastfeed may be an important strategy to increase breastfeeding rates and duration.

The negative effects of early weaning on children and mothers remain a significant public health concern. An analysis of data from the 2001 Australian National Health Survey found that fewer than 50% of infants were receiving breastfeeding milk at six months [22], which is considerably lower than the 80% figure recommended by the latest Australian Dietary Guideline for Children and Adolescents [3,22]. In addition, very few Australian infants are being exclusively breastfed for the recommended six months [8]. The lack of knowledge about the recommended duration of exclusive breastfeeding among first-time mothers in our study is likely to have contributed to a reduced number of mothers who intended to breastfeed exclusively for six months.

To date, most studies on breastfeeding awareness have focused on the health benefits of breastfeeding and infant feeding practices [4,7,9,10]. Few studies have looked into whether mothers actually understand the recommendations and what the recommendations mean to them. In interviews with some of the mothers participating in our study (data not presented), we found that they had limited understanding of the term "exclusive breastfeeding." In addition they expressed concerns about the quantity and quality of breast milk, and whether breast milk alone would be sufficient for their infant for six months. These findings were consistent with other studies suggesting that the most common reason cited by mothers for stopping breastfeeding was that the baby was unsettled, a behaviour often interpreted by mothers as indicating an insufficient milk supply [23]. This perception of insufficient supply appears to be due to a lack of information or lack of confidence regarding the normal process of lactation [24].

This study provides empirical evidence linking mothers' awareness of the breastfeeding recommendations and their intention to meet the recommendation. While our study had a relatively large sample of 409 first-time mothers, its generalisability is limited due to the locality of the study area. Southwest Sydney is the most socially and economically disadvantaged area of metropolitan Sydney [20]. We acknowledge the need to exercise caution in making assertions of the causal relationship between breastfeeding awareness and the intention to breastfeed based on a cross-sectional survey of this kind. Further studies are required to establish more definitively whether being aware of the breastfeeding recommendations actually improves intention to breastfeed and increases breastfeeding duration. In addition, whether model of pregnancy care plays a role in breastfeeding awareness and intention needs to be explored further.

Conclusion

Potentially, our findings have a number of important policy implications for breastfeeding interventions. Efforts to encourage mothers to meet the recommendations should focus on improving mothers' knowledge and understanding of the recommendations, and address the concerns expressed by mothers about the quantity and quality of breast milk for the recommended duration. The effectiveness of targeted health promotion programs needs to be tested, particularly among young, unemployed and less educated mothers. There is a need to improve their perceptions, attitudes and knowledge, and also to change social norms in relation to breastfeeding practices.

In addition, appropriate public health policies to help mothers to breastfeed to at least six months, and to remove the barriers to breastfeeding, will be required to meet the WHO recommendations. The Australian Government's proposed 18 weeks paid maternity leave is a good start, but falls short of the required length, and falls particularly short of some Scandinavian and European countries' maternity leave entitlements of 50 to 64 weeks. For women who need to or choose to return earlier to work, the workplace needs to be able to provide child care and to facilitate breastfeeding on demand. Strategies recommended in the NSW Breastfeeding Policy Directive [25] provide examples for worksites on how to promote, protect and support breastfeeding in the community and amongst staff. These structural changes, along with health promotion programs which include changing social attitudes to breastfeeding, may improve the capacity of the most disadvantaged women to consider breastfeeding longer and more exclusively.

Competing Interests

The authors declare that they have no competing interests.

Authors' Contributions

LMW, LB, CR and GA conceived the HBT, and contributed to the development of the trial and the procurement of the funding.

In this study, LMW undertook literature review, data analysis and interpretation and wrote the original draft. JS provided advice on data analysis. LB, CR, GA and JS made significant comments on the draft. All authors have read and approved the final manuscript.

Acknowledgements

This is part of the Healthy Beginnings Trial funded by the Australian National Health and Medical Research Council (ID number: 393112). We sincerely thank the Associate Investigators, Prof. Anita Bundy, Dr Lynn Kemp and Dr Vicki Flood and the members of the steering committee and working group for their advice and support. We wish to thank all the families for their participation in this study. We also thank members of the project team including Karen Wardle, Carol Davidson; Cynthia Holbeck; Dean Murphy; Lynne Ireland, Brooke Dailey, Kim Caines and Angela Balafas. In addition, we wish to thank James Kite, and Therese Carroll for their support in setting up the database and Hui Lan Xu for assisting data entry and analysis.

References

1. Horta BL, Bahl R, Martinés JC, Victora CG: Evidence on the long-term effects of breastfeeding: Systematic reviews andmeta-analysis. World Health Organization Geneva, Switzerland; 2007.

2. World Health Organization: Global Strategy for Infant and Young Child Feeding. Report by the Secretariat (WHA55 A55/15, paragraph 10, page 5) April 2002, Geneva, Switzerland

3. National Health and Medical Research Council: Dietary Guidelines for Children and Adolescents in Australia: Incorporating the Infant Feeding Guidelines for Health Workers. Canberra, Australia: Commonwealth Department of Health and Ageing; 2003.

4. Allen J, Hector D: Benefits of Breastfeeding. New South Wales Public Health Bulletin 2005, 16(4):42–46.

5. American Academy of Pediatrics: Policy statement: Breastfeeding and the use of human milk. Pediatrics 2005, 115:496–506.

6. Amir LH, Donath SM: Socioeconomic status and rates of breastfeeding in Australia: evidence from three recent national health surveys. The Medical Journal of Australia 2008, 189(5):254–6.

7. Hector D, Webb K, Lyner S: Describing breastfeeding practices in New South Wales using data from the NSW Child Health Survey, 2001. New South Wales Public Health Bulletin 2005, 16(3-4):47–51.

8. Garden F, Hector D, Eyeson-Annan M, Webb K: Breastfeeding in New South Wales: Population Health Survey 2003-2004. Sydney; NSW Centre for Public Health Nutrition, University of Sydney, and Population Health Division, NSW Department of Health; 2007.

9. Chambers JA, McInnes RJ, Hoddinott P, Alder EM: A systematic review of measures assessing mothers' knowledge, attitudes, confidence and satisfaction towards breastfeeding. Breastfeeding Review 2007, 15(3):17–25.

10. Papinczak TA, Turner CT: An analysis of personal and social factors influencing initiation and duration of breastfeeding in a large Queensland maternity hospital. Breastfeeding Review 2000, 8(1):25–33.

11. Kong SKF, Lee DTE: Factors influencing decision to breastfeed. Journal of Advanced Nursing 2004, 46:369–379.

12. Chezem J, Friesen C, Boettcher J: Breastfeeding knowledge, breastfeeding confidence, and infant feeding plans: effects on actual feeding practices. Journal of Obstetric, Gynecologic, & Neonatal Nursing 2003, 32(1):40–47.

13. Riva E, Banderali G, Agostoni C, Silano M, Radaelli G, Giovannini M: Factors associated with initiation and duration of breastfeeding in Italy. Acta Paediatrica 1999, 88:411–415.

14. Donath S, Amir LH: Rates of breastfeeding in Australia by state and socioeconomic status: Evidence from the 1995 National Health Survey. Journal Paediatrics and Child Health 2000, 36(2):164–168.

15. Britton C, McCormick FM, Renfrew MJ, Wade A, King SE: Support for breastfeeding mothers. Cochrane Database of Systematic Reviews 2007, (1):CD001141. DOI: 10.1002/14651858.CD001141.pub3

16. Forster DA, McLachlan HL, Lumley J: Factors associated with breastfeeding at six months postpartum in a group of Australian women. International Breastfeeding Journal 2006, 1:18.

17. Donath S, Amir LH, ALSPAC Study Team: Relationship between prenatal infant feeding intention and initiation and duration of breastfeeding: a cohort study. Acta Paediatrica 2003, 92(3):352–356.

18. Rempel LA: Factors influencing the breastfeeding decisions of long-term breastfeeders. Journal of Human Lactation 2004, 20:306–317.

19. Wen LM, Baur LA, Rissel C, Wardle K, Alperstein G, Simpson JM: Early intervention of multiple home visits to prevent childhood obesity in a disadvantaged population: a home-based randomised controlled trial (Healthy Beginnings Trial). BMC Public Health 2007, 7:76.

20. Centre for Epidemiology and Research, NSW Department of Health: New South Wales Adult Health Survey 2003. NSW Public Health Bulletin 2004., 15(S-4):

21. StataCorp: Stata Statistical Software: Release 10. Colleague Station, TX: StataCorp LP; 2007.

22. Donath SM, Amir LH: Breastfeeding and the introduction of solids in Australian infants: data from the 2001 National Health Survey. Australian & New Zealand Journal of Public Health 2005, 29(2):171–175.

23. Binns CW, Scott JA: Breastfeeding: reasons for starting, reasons for stopping and problems along the way. Breastfeeding Review 2002, 10:13–19.

24. Powers NG: Slow weight gain and low milk supply in the breastfeeding dyad. Clinics in Perinatology 1999, 26(2):399–430.

25. NSW Health: Breastfeeding in NSW: Promotion, Protection and Support (Policy Directive). [http://www.health.nsw.gov.au/policies/pd/2006/PD2006_012.html] Doc No.: PD2006_012, NSW Health 2006.

An Exploration of How Clinician Attitudes and Beliefs Influence the Implementation of Lifestyle Risk Factor Management in Primary Healthcare: A Grounded Theory Study

Rachel A. Laws, Lynn A. Kemp, Mark F. Harris, Gawaine Powell Davies, Anna M. Williams and Rosslyn Eames-Brown

ABSTRACT

Background

Despite the effectiveness of brief lifestyle intervention delivered in primary healthcare (PHC), implementation in routine practice remains suboptimal.

Beliefs and attitudes have been shown to be associated with risk factor management practices, but little is known about the process by which clinicians' perceptions shape implementation. This study aims to describe a theoretical model to understand how clinicians' perceptions shape the implementation of lifestyle risk factor management in routine practice. The implications of the model for enhancing practices will also be discussed.

Methods

The study analysed data collected as part of a larger feasibility project of risk factor management in three community health teams in New South Wales (NSW), Australia. This included journal notes kept through the implementation of the project, and interviews with 48 participants comprising 23 clinicians (including community nurses, allied health practitioners and an Aboriginal health worker), five managers, and two project officers. Data were analysed using grounded theory principles of open, focused, and theoretical coding and constant comparative techniques to construct a model grounded in the data.

Results

The model suggests that implementation reflects both clinician beliefs about whether they should (commitment) and can (capacity) address lifestyle issues. Commitment represents the priority placed on risk factor management and reflects beliefs about role responsibility congruence, client receptiveness, and the likely impact of intervening. Clinician beliefs about their capacity for risk factor management reflect their views about self-efficacy, role support, and the fit between risk factor management ways of working. The model suggests that clinicians formulate different expectations and intentions about how they will intervene based on these beliefs about commitment and capacity and their philosophical views about appropriate ways to intervene. These expectations then provide a cognitive framework guiding their risk factor management practices. Finally, clinicians' appraisal of the overall benefits versus costs of addressing lifestyle issues acts to positively or negatively reinforce their commitment to implementing these practices.

Conclusion

The model extends previous research by outlining a process by which clinicians' perceptions shape implementation of lifestyle risk factor management in routine practice. This provides new insights to inform the development of effective strategies to improve such practices.

Background

Lifestyle risk factors such as smoking, poor nutrition, excessive alcohol consumption, and physical inactivity are a major cause of preventable mortality, morbidity, and impaired functioning [1,2]. The World Health Organisation estimates that 80% of cardiovascular disease, 90% of type 2 diabetes, and 30% of all cancers could be prevented if lifestyle risk factors were eliminated [1]. Primary healthcare (PHC) has been recognised as an appropriate setting for individual intervention to reduce behavioural risk factors because of the accessibility, continuity, and comprehensiveness of the care provided [3]. A growing body of evidence suggests that brief lifestyle interventions delivered in PHC are effective [4-8], and the 5A's principle of brief intervention (ask, assess, advise, assist, and arrange) has been widely endorsed in preventive care guidelines [9-12].

Despite this, implementation of risk factor management in routine practice remains low. Screening for lifestyle risk factors does not occur routinely, and only a fraction of 'at risk' patients receive any intervention in PHC [13-16]. Furthermore, studies suggest that when lifestyle intervention is provided it tends to be limited to asking and giving advice on the health risks of the behaviour rather than providing assistance, referral, or follow up needed to support behaviour change [17,18]. The findings of intervention studies aimed at enhancing risk factor management practices have been mixed and often disappointing [19-22]. These studies have used a range of intervention strategies; however, they provide little information about the theoretical or conceptual basis for their choice of intervention and limited contextual data. This suggests that to improve practices a better conceptual understanding of the factors impacting on the implementation of lifestyle risk factor management in routine PHC is required.

Research examining lifestyle risk factor management practices has consisted predominantly of descriptive studies of barriers or enablers, or cross sectional studies of self-reported practices conducted in general practice. These studies have consistently identified the importance of clinician beliefs, including perceptions about role congruence [23-26], self-efficacy [18,27-29], beliefs about effectiveness of interventions [24,25,30-33] and patient motivation [23,34], concern regarding client acceptance [23-25], as well as personal lifestyle behaviours [24,35,36]. Few studies have been conducted beyond general practitioner (GP) PHC providers. Studies among PHC nurses, including community nurses, and registered and licensed practical nurses in USA and Finland, have also reported the importance of clinician beliefs and attitudes, mirroring the findings in general practice [36-39].

Our previous research suggests that those who frequently address risk factors with their patients have different beliefs and attitudes from those who do so less frequently [40]. However, as cross-sectional studies these can provide only limited insight into the way clinician perceptions shape risk factor management practices, and the impact of structural or contextual factors on this. A better conceptual understanding of how clinician beliefs and attitudes influence the implementation of risk factor management in PHC is required to guide the development of effective strategies to improve practice.

This study builds on our previous cross-sectional study [40] and aims to: describe a theoretical model for understanding how clinician perceptions shape their implementation of lifestyle risk factor management in routine practice; and discuss the implications of the model for developing interventions to improve these practices.

Methods

This study used grounded theory principles, a research method designed to generate a theoretical explanation of a social phenomenon that is derived from (grounded in) empirical data rather than from a preconceived conceptual framework [41], and therefore well suited to understanding process from the perspective of participants [42]. The approach to grounded theory adopted in this study was informed by a constructionist perspective [43] which assumes that neither data nor theories are discovered but constructed based on shared experiences between researchers and participants [43]. Hence, the model produced is a construction of reality offering plausible accounts and explanations rather than verifiable knowledge.

Study Setting and Context

This research was part of a larger feasibility project, the details of which have been reported elsewhere [44,45]. In brief, the project aimed to develop and test approaches to integrating the management of lifestyle risk factors into routine care among PHC providers outside of the general practice setting. It involved three community health teams from two Area Health Services (AHS) in the state of New South Wales (NSW), Australia. In NSW, AHS are responsible for providing all hospital- and community-based healthcare apart from general practice and PHC services for specific population groups such as Aboriginal and Torres Strait Islanders. Community health services are the second largest provider of publicly funded PHC services to the general population after GPs [46].

All eight AHS in NSW were invited to express interest in participating in the study and to nominate suitable teams. A total of three community health teams were selected from two of three AHS who expressed interest. Selection was based on the capacity of the team to be involved and the relevance of risk factor management to the type of service provided and healthcare context. Teams were also selected to maximise the variability in team characteristics including provider type, team location (co-located or not), geographical locality, management structures, and health system context.

Team one (n = 35) was a generalist community nursing team with both enrolled and registered generalist community nurses, located in a metropolitan area. Team two (n = 16) was a co-located multi-disciplinary community health team from a rural area, while team three (n = 10) consisted of PHC nurses, Aboriginal health workers, and allied health practitioners providing PHC services to rural and remote communities that generally did not have access to other health services such as a GP. In each of the teams, a baseline needs assessment was conducted to determine current lifestyle risk factor management practices, factors shaping practices, and supports required to improve practices. This needs assessment then informed the development and implementation of a capacity building intervention to enhance practices which was tailored to the needs of each team. Following a six-month implementation period further data was collected to determine changes in practices and factors influencing uptake of practices.

Data Sources and Collection Procedures

This study utilised two sources of data collected as part of the larger feasibility project: semi-structured interviews with participants prior to and six months following the capacity building intervention undertaken with each team; and project manager journal of reflections and observations recorded throughout the feasibility project.

As part of the feasibility project, semi-structured interviews were conducted with a purposeful sample of participants across the three teams at baseline (n = 29) and six months following the team capacity building intervention (n = 30). At baseline, the aim was to interview a sample of clinicians from across the three teams who varied in profession and role (enrolled and registered nurses, allied health staff, Aboriginal health workers and managers), experience, and geographical location. The same participants were invited to take part in an interview post-intervention (where possible) to provide comparative data on the same individuals over time. A concerted effort was also made to identify and approach to take part in an interview those who felt less positive about the project and risk factor management in general. These clinicians were identified through response on a post-intervention survey and through discussions with managers and project officers responsible for local implementation.

Full details of the data collection procedures for the qualitative interviews have been reported elsewhere [40,45]. In brief, the baseline interviews were conducted by the project manager (lead author RL) and covered issues related to barriers, enablers, and capacity to undertake risk factor management from the perspective of both clinicians and managers (Table 1). Following the project, an evaluation officer (REB) not involved in implementing the team intervention conducted interviews to explore participants' experience of attempting to integrate risk factor management into routine work (Table 1). Interviews at baseline and post-intervention lasted between 20 and 75 minutes, and were tape recorded with participants' permission and transcribed verbatim. The project manager (lead author RL) also kept a journal throughout the two-year project to record reflections and observations following interaction with clinicians and managers during field visits and following participant interviews. All journal notes were typed and included in the analysis for this study.

Table 1. Topic guide for baseline and post-intervention interviews conducted as part of the feasibility project

Baseline Interviews	Post-Intervention Interviews
• Overview of job role	• General impressions of the project
• How addressing SNAP risk factors fits with the job role/core business of team or service[2]	• Case example—last client with a risk factor[1]
• Approach to addressing SNAP risk factors (client case example)[1]	• Feasibility of risk factor screening/intervention
• Work priority to address SNAP risk factors[1]	• Barriers/enablers risk factor screening/intervention
• Confidence to address SNAP risk factors[1]	• Case example—comfortable to address[1]
• Barriers and enablers to addressing SNAP risk factors in routine work	• Case example—not comfortable to address[1]
• Support and resources required to address SNAP risk factors in routine work/strengthen team capacity to address risk factors[2]	• Perceived effectiveness of intervening[1]
• Opinion on strength of local referral networks and programs to support risk factor management[2]	• Congruence with core business of the team and organisation[2]
• Opinion on team climate and any competing priorities in implementing the project[2]	• Process of project implementation (degree of consultation and model adaptation to suit team)[2]
	• Change in approach to addressing SNAP risk factors
	• Views about continuation of risk factor management as part of professional role[1]/team or service[2]
	• Support required for continuation of risk factor management practices in professional role[1]/team or service[2]
	• Project benefits (personal and professional[1]/team or service[2])

SNAP: Smoking, nutrition, alcohol and physical activity
[1]Team and service managers only
[2]Team/service managers and project officers only

Data Analysis and Model Development

Developing the model involved two main stages of analysis. First, a preliminary model was developed by analysing a purposeful selection of baseline interviews (n = 18) of participants who also participated in an interview following the project, allowing for comparison over time. Analysis at this stage involved open and focused coding to identify key theoretical categories and ideas about how these were related [47]. From this process, a preliminary model was constructed and compared to relevant theories in the literature in order to identify 'conceptual gaps,' heightening the researcher's theoretical sensitivity [48,49].

The second stage of analysis involved refining the preliminary model through analysis of additional interviews (n = 30) and the project managers' journal notes. In line with grounded theory principles [41,50], 10 interviews were theoretically sampled from the existing interview dataset. A sampling frame was devised (Table 2) in order to identify those with a diverse range of attitudes and practices relevant to the evolving model. Clinician response on a risk factor management survey undertaken at baseline and post-intervention was used to identify clinicians meeting the sampling criteria (details of the survey reported elsewhere [40]). A further 20 interviews were purposefully selected including post-intervention interviews for those who had participated in an interview at baseline (n = 18) and interviews with project officers (n = 2) involved in implementing the project locally. Analysis at this stage involved assessing how well the focused codes developed in the preliminary model fitted the new data. This process resulted in the revision of some categories (for example, to include additional properties and dimensions) and the development of additional categories to reflect the data. Baseline data was then recoded using the new and revised categories to ensure the conceptual fit with the data. Theoretical coding was then used to specify the possible relationships between the categories developed during focused coding to construct a coherent analytical story [41,42,47]. Preliminary ideas about relationships were tested by going back to the data in accordance with grounded theory principles of moving between induction and deduction in the development of theory [42].

Table 2. Criteria used to theoretically sample interviews to include in the analysis

Factors related to key categories in the baseline model
Clinicians who scored low[1] or high[2] on the following attitude items completed as part of a survey at baseline and/or post-intervention:
- The acceptability of raising risk factor issues with clients
- Perceived work priority
- Perceived effectiveness of addressing lifestyle issues
- Confidence in assessing and managing lifestyle risk factors
- Confidence in applying behaviour change
- Perceived accessibility of support services

Other criteria included
- Clinician types not included in the baseline analysis
- Clinicians reporting change[3] in confidence and/or attitudes from baseline to post-intervention
- Clinicians and managers who have recently joined the team (last six months)

Clinician screening and intervention practices
- Clinicians who had low or high levels of self reported screening for lifestyle risk factors at baseline and/or post-intervention[4]
- Clinicians who had low or high levels of self reported intervention for lifestyle risk factors at baseline and/or post-intervention[5]
- Clinicians reporting a change[3] in screening and or intervention practices from baseline to post-intervention

[1] Low defined as scores in the clinician risk factor survey in the lowest quartile for those participating in an interview
[2] High defined as scores in the clinician risk factor survey in the highest quartile for those participating in an interview
[3] Change defined as scores increasing from lowest to highest quartile or highest to lowest quartile (baseline to post-intervention)
[4] High screening practices = mean screening score (across risk factors) in the highest quartile for those participating in an interview, low screening practices = mean screening score in the lowest quartile for those participating in an interview
[5] High intervener = high frequency of intervention for three or more risk factors and/or high intensity intervention (across risk factors), low intensity intervener = low frequency of intervention for three or more risk factors and/or low intensity intervention (across risk factors).

Throughout the analysis process, constant comparative techniques were used to assist in uncovering the properties and dimensions of each category. This involved comparing data within the same coding group, making comparisons between different clinicians and between the same clinician over time. In line with Strauss and Corbin's [42] notion of axial coding, attention was paid to identifying and comparing the conditions giving rise to an issue, the context in which it was embedded, the strategies used by clinicians to manage this, and the consequences for clinicians beliefs and practices. Insights gained were recorded in the form of memos throughout the analysis process. NVivo 7.0 software [51] was used to attach codes to text, record memos, and diagrams, as well as facilitate the retrieval of data.

One member of the research team (RL) undertook the analysis. To avoid the researchers' views being 'imposed' on the data, RL documented assumptions prior to analysis and kept an audit trail to document coding decisions, which included extensive use of participant quotes to justify the approach taken [52]. A conscious decision was made not to use member checking, a process of cross-checking findings and conclusions with participants. As the purpose of the analysis was to code all responses and organise into a new higher order theoretical model, it was not expected that participants would be able to recognise their individual contributions or concerns. It was therefore not appropriate to seek 'validation' from individual participants. Instead, a number of other techniques were used to ensure interpretations were grounded in the data. These included the use of constant comparisons, memo writing, extensive use of participant quotes, and discussing coding frameworks and preliminary theoretical ideas with two other members of the research team (MH and LK) for the purpose of gaining other perspectives and challenging assumptions rather than to reach agreement.

Ethics

The project was approved by the UNSW Human Research Ethics Committee (HREC) and the HREC in each AHS.

Results

The final sample in this study included 48 interviews with 23 clinicians, three team managers, two senior community health managers, and two project officers. Fourteen clinicians and four managers were interviewed twice, at the beginning and end of the project. The sample included generalist community nurses, child and family nurses, a range of allied health providers, and one Aboriginal health worker. All were female, with a wide range of professional experience. The interview sample included in this study was broadly representative of clinicians from the three teams (Table 3). However, allied health practitioners and child and family nurses from team two were over-represented and males under-represented in the interview sample. This reflected the purposeful and theoretical sample techniques that aimed to include a diverse range of clinician types and those with varying levels of attitudes and practices related to the management of lifestyle risk factors.

Table 3. Characteristics of clinicians included in the interview sample compared to all clinicians

	Clinician interviews included in analysis (n = 23)	All clinicians[1] (n = 61)
Age Category	No. (%)	No. (%), n = 57
18 to 24 years	2 (8.7)	2 (3.5)
25 to 34 years	2 (8.7)	6 (10.5)
35 to 44 years	8 (34.8)	16 (28.1)
45 to 54 years	8 (34.8)	26 (45.6)
55 to 64 years	3 (13.0)	7 (12.3)
Clinician experience	Mean (std), range	Mean (std), range n = 60
Years in profession	21.0 (11.4), 1-35.0	21.6 (11.0), 1-46.0
Years in community health	8.4 (8.1), 0.5-30.0	10.5 (7.8), 0.5-30.0
Years in team	6.8 (6.5), 0.5-20.0	6.5 (6.1), 0.5-22
Gender	No. (%)	No. (%), n = 60
Male	0 (0.0)	3 (5.0)
Female	23 (100.0)	57 (95.0)
Employment	No. (%)	No. (%), n = 55
Part time	12 (52.2)	26 (47.3)
Full time	11 (47.8)	29 (52.7)
Clinician type	No. (%)	No. (%), n = 60
Generalist community nurse (registered nurse)	12 (52.2)	37 (61.7)
Generalist community nurse (enrolled nurse)	3 (13.0)	11 (18.3)
Child and family nurse	2 (8.7)	2 (3.3)
Allied health practitioner	5 (21.7)	8 (13.3)
Aboriginal health worker	1 (4.3)	2 (3.3)
Team	No. (%)	No. (%), n = 61
Team one	9 (39.1)	35 (57.4)
Team two	10 (43.5)	16 (26.2)
Team three	4 (17.4)	10 (16.4)

[1]Demographic information collected at baseline as part of clinician survey. Missing data: age n = 4; gender n = 1, employment n = 6, clinician type n = 1.

Model Overview

The theoretical model is shown in Figure 1. It suggests that clinician perceptions shape their risk factor management practices through the process of 'practice justification.' This involves justifying risk factor management practices as a legitimate, 'doable,' and worthwhile component of the role. This process consists of five main interrelated factors:

1. Developing commitment (Should I address lifestyle issues?)
2. Assessing capacity (How can I address lifestyle issues?)

3. Formulating intervention role expectations/intentions (How will I intervene?)
4. Implementing risk factor management practices
5. Weighing up benefits and costs of practice (Is it worth it?)

Each of these steps in the model is described below.

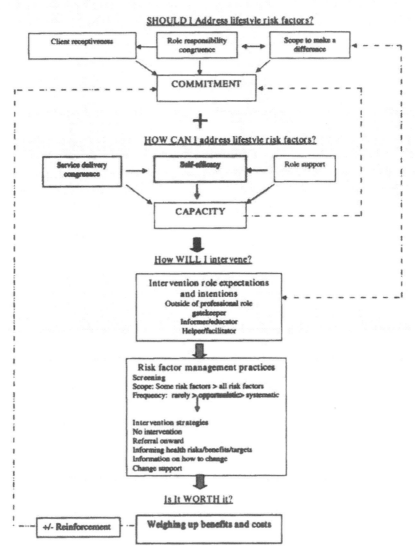

Figure 1. The practice justification process: A model of how clinician perceptions shape their risk factor management practices.

Developing Commitment — should i Address Lifestyle Issues?

First, 'commitment' represents the priority or importance placed on risk factor management in the role, influencing 'if and when' clinicians address lifestyle issues, the amount of time they are willing to invest, and the scope of their practice (type of risk factors addressed and frequency in which this occurred). Commitment in turn appeared to be shaped by three main factors, as outlined in Figure 1: role responsibility congruence, perceptions of client receptiveness, and beliefs about the 'scope to make a difference.'

Clinicians expressed a diversity of views about how addressing lifestyle issues fitted with their role responsibilities. For some, it was simply an assessment task to 'tick off' before getting on with the job of looking after the clients; for others the relevance varied, depending on the clients presenting problem. In contrast, other clinicians saw risk factor management as an integral component of their role in providing holistic PHC, as articulated by this clinician:

'My approach is...holistic health and wellness...so ultimately what I'm looking for is information to assist people being totally healthy and well...So continuing to assess and support lifestyle changes, yeah I do believe it should continue to be part of our role.' (Clinician 23)

Overall, the broader clinicians' perspective of the relevance of lifestyle issues to their role, the more they were willing to invest time in addressing them. These views tended to reflect the model of service delivery adopted by the team/service in which they worked and clinicians' discipline and training. For example, only generalist community nurses or PHC nurses considered addressing lifestyle issues as part of their role in providing holistic care, while the relevance for allied health practitioners depended on the link between risk factor issues and the clients presenting problem. For those with a counselling role (such as psychologist and social workers), screening for lifestyle issues was considered to be in conflict with their client-centered approach, and they considered it only appropriate to address risk factor issues opportunistically when relevant to the clients concerns:

'I think for the nurses, it's very feasible because the nurses tend to be holistic and cover absolutely everything, and I think for the allied health, its still quite feasible, perhaps not all the [risk] factors like the nurses...for the... counselling type people, I think it's been harder for them to do it just because they have such a 'let the client take the direction focus.' (Project Officer 2)

In addition to clinicians' intrinsic sense of their professional responsibility to address lifestyle issues, their perception of client receptiveness was an important driver of their commitment to broach these topics. Clinicians who reported that clients were receptive to them asking about lifestyle risk factors expressed confidence and commitment to raising these issues. Lifestyle risk factors were considered easier to raise, and clients most receptive, when the client was being seen for a preventive or PHC issue, and when the clinician had ongoing contact with the client in a case management role. Some clinicians considered lifestyle issues more difficult to raise when seeing clients in their own home due to clients control over the care agenda and the clinicians assumed role of a 'guest' who does not want to offend their 'host.' Clinicians also deemed clients to be less receptive when they had other pressing problems, or when they were unreceptive to the care process in general. When clinicians expressed concern about client receptiveness, they discussed feeling less confident and committed to broaching lifestyle topics because of the implications a negative reaction might have for their own safety and/or their relationship with the client, as illustrated in this quote:

'You go in there as a single nurse on your own, and if you don't approach the subjects in the right way, you could end up in a little bit of an uncomfortable situation...' (Clinician 18)

Finally, clinician commitment not only reflected their beliefs about their professional responsibility and client receptiveness, but the extent to which they believed that intervening could have a positive impact (labelled 'scope to make a difference': Figure 1). Clinicians were doubtful and sometimes openly pessimistic about whether intervening would make a difference when they:

1. considered the benefits of intervening only at the individual level and in terms of primary prevention of disease;

2. did not see a role for themselves in motivating clients to change behaviour: hence a lack of client motivation was considered a major barrier in certain groups of clients (e.g., older clients, those with other pressing problems);

3. judged the effectiveness of intervention in terms of the number of clients achieving the desired behavioural targets;

4. attended clients for one off, or short term services where there was limited opportunity to build rapport or follow up outcomes achieved.

In these circumstances addressing lifestyle issues was considered to be of limited use and hence, commitment to doing this was low, as argued by this clinician:

'If, unfortunately intellectually they can't...take those issues on board, really nothing much you say can alter lifestyle patterns that are from birth. So, I tend...look at what I can change and try to change it, and if I don't think I can, then I just move around it.' (Clinician 4)

In contrast, clinicians were more likely to identify greater scope to make a difference when they:

1. took a broader view of the benefits of intervening beyond the individual, and for the purposes of primary through to tertiary prevention and maintenance of quality of life;

2. viewed their role as facilitators of change: hence a lack of client motivation was not considered a deterrent but part of the process;

3. judged the effectiveness of their intervention in terms of the process of change rather than achieving behavioural targets, considering their intervention as one of many which may impact on the prevalence of lifestyle risk factors at the population level.

Not surprisingly, these clinicians also adopted a broader view of their role responsibilities beyond the presenting issue to providing PHC services to families and communities. Clinicians' belief about their own ability and capacity to effect change was also important in shaping their perceptions about the likely impact of intervening, as discussed below.

Assessing Capacity—How can i Address Lifestyle Issues?

Clinicians' risk factor management practices not only reflected their beliefs about whether they 'should' address lifestyle issues (commitment) but also their beliefs about how they 'can' address lifestyle issues (capacity). Three main components of capacity were identified to be important in shaping practices (Figure 1): self-efficacy, role support, and service delivery congruence.

First, in order for clinicians to feel confident addressing lifestyle issues, they needed to believe that they had the ability to do so, based on internal factors, such as knowledge, skills, experience, and their own lifestyle habits. This has been labelled 'self-efficacy' in the model and appeared to be important in determining the type of intervention offered, as discussed by this clinician:

'...I suppose maybe it's based on how comfortable...or personally confident I feel about offering anything... I certainly would refer to the relevant person but not deal with it specifically myself.' (Clinician 6)

To feel confident offering an intervention themselves, clinicians discussed the importance of having an understanding of various intervention strategies either through their own experience of lifestyle change or through their work with clients. However, they also recognised a need for a sound grasp of behaviour change skills, such as motivational interviewing, if they were to move beyond providing information and advice to facilitating behaviour change

Perceptions about capacity not only reflected clinicians' confidence about their own abilities but also external factors such as access to support mechanisms, labelled 'role support' (Figure 1). This included decision support tools (such as screening tools), ongoing training, client education materials, collegial support, and access to referral services for clients. These mechanisms appeared to increase clinicians' confidence to intervene by enhancing perceptions of self-efficacy, and by providing 'back up' support and 'something tangible' to offer clients:

'Now they have somewhere they can refer them to because before [the project] even if they wanted to address it, it was like, 'oh well, what's the point, where can I refer them to.'.. but now that they know that there is actually something, I think it makes a big difference.' (Manager 5)

Data analysis suggests that access to these support mechanisms is dependent on having wider system level support for risk factor management at the service and organisational level, including good linkages with support services.

Finally, the work environment was important in shaping perceptions about capacity, in particular the fit between risk factor management and ways of working (labelled 'service delivery congruence': Figure 1). As part of the project, teams were consulted about the most appropriate way for them to address lifestyle issues, given their current way of working. This consultation process was identified as an important moderator to developing approaches that fitted with the mechanics of everyday practice. At the macro-level, the extent to which risk factor management was seen to fit with the model of service delivery was also important in shaping clinician's beliefs about the opportunities they had for implementation. For example, all community nurses interviewed in team one identified the focus on providing post-acute care as limiting the time available for health promotion activities. Some allied health providers also questioned their capacity to address lifestyle issues peripheral to the reason for referral, given that they were solo practitioners with long waiting lists and limited ongoing contact with clients. In contrast, team three considered risk factor management as central to delivering PHC services to rural and remote communities with a focus on early intervention and prevention, as summed up by this participant:

'...we have chronic disease prevention and early intervention as one of the five priority health areas...so it [risk factor management] fits really well into...our core business.' (Team 3)

Formulating Intervention Role Expectations/Intentions – How will i Intervene?

Analysis of the data suggests that clinicians formulate different expectations and intentions about how they will intervene based on their beliefs about commitment and capacity and their philosophical views about appropriate ways to intervene (Table 4, Figure 1). Philosophical views appeared to reflect a diversity of beliefs about the determinants of lifestyle behaviours and how they should be best managed. Role expectations ranged from seeing lifestyle risk factor management as completely outside of the professional role and best managed through population health approaches, to those who considered they had an important role to play in facilitating behaviour change by providing tailored support strategies (Table 4). These role expectations and intentions appeared to act as a cognitive framework or mindset shaping clinicians' intervention practices.

Table 4. Intervention role expectations and intentions: Description and illustrative quotes

Intervention role expectations/intentions	Philosophical views about appropriate ways to intervene	Illustrative quotes
Expectations–Outside of Professional Role Intervention considered outside of the professional role, best addressed through population health approaches **Intentions:** Do not intervene to address lifestyle issues	**Population Health Perspective:** Lifestyle behaviours best tackled through addressing underlying determinants of risk taking behaviour	'It wouldn't be us that would be able to take that extra work on…It'd have to be like those ones that do the programs like population [health]Like you people and all that that get funded for these things would have to carry it further.' (Clinician 22)
Expectations–Gatekeeper Intervention considered outside of scope of professional expertise and job role, best addressed by qualified experts. **Intentions:** Refer clients onwards to qualified experts/ specialist service	**Medical perspective:** Lifestyle behaviours are complex and require specialist input from qualified experts	'It's not my job to get people to quit smoking/ they want to quit smoking I would give them the quit line number/ don't have…those skills…if I was a drug and alcohol worker it'd be a different story, but I'm not.' (Clinician 15)
Expectations–Informer and educator Ensure client has sufficient information to make an informed choice about lifestyle behaviour. Can only provide intervention to those willing to change. **Intentions** Provide information on health risks/benefits of lifestyle risk factors to all clients. Provide additional assistance to motivated clients.	**Individual perspective (individual autonomy and self empowerment):** Lifestyle behaviours are personal choices that people make and as such should be respected. Individuals need to take responsibility for change	'I really leave it up to them—it's their decision what they're going to do, but at least I can give them the information so they can reach a decision whether to keep on smoking or stop.' (Clinician 7)
Expectations– Helper or facilitator Help move clients towards change over time by acting as a facilitator. Synergistic role with other providers and population health approaches. **Intentions:** Facilitate clients to change their behaviour through providing tailored support strategies.	**Socio-ecological perspective** Lifestyle habits are complex behaviours influenced by a range of social and environmental factors. Multiple interventions required at individual and population level to effect change.	'If everybody got together and said these risk factors well then people are going to think …and obviously it's working with the…TV advertising…our smoking rates are going down…' (Clinician 14)

Risk Factor Management Practices

Clinicians' risk factor management practices varied according to the approach adopted for assessing lifestyle risk factors (opportunistic versus systematic), the type of risk factors addressed (all or selective risk factors), and the range of intervention strategies used (Figure 1). Practices varied between clinicians and also by the risk factor being addressed (for some clinicians). These variations can be best understood in terms of the key model categories of commitment, capacity, and intervention role expectations and intentions.

A small number of clinicians reported infrequently broaching lifestyle issues. This reflected both a lack of commitment and capacity. First, lifestyle risk factors were not generally considered relevant to the clients presenting problem, and thus clients were unlikely to be receptive to discussing these issues. Screening for lifestyle risk factors was also not part of their usual work process, they reported having limited opportunities to intervene and they lacked the necessary knowledge, skills, and access to support tools/resources. Clinicians who reported adopting an opportunistic approach to asking about selective risk factors with particular clients did not routinely ask about lifestyle issues as part of existing work processes. Hence, they took an opportunistic approach to broaching these topics when the lifestyle issues were considered relevant to the clients presenting problems, and when the client was likely to be interested and able to make lifestyle changes. In contrast, those clinicians who reported using a systematic approach to asking about most lifestyle issues with the majority of their clients took a broader view of the relevance of lifestyle risk factors to their role and/or asking about lifestyle issues was integrated into the standard assessment process.

Once risk factors were identified, clinicians' intervention practices ranged from providing no intervention (one clinician) to providing personalised support for lifestyle change tailored to the clients' situation (Figure 1, Table 5). Intervention strategies differed in terms of the time, knowledge, and skill required to deliver

Table 5. Intervention strategies: Illustrative quotes

No intervention
'I would never discuss the interventions. We never got that far...we do not have clients that these things are practical for.' (Clinician 22)

Referral onward
'I think the most I have done is referred someone to Quitline but in terms...of doing anything I haven't really done a lot.' (Clinician 5)

Informing of health risks/benefits and lifestyle targets
'Recently, I saw a gentleman...probably early 70s who has obviously been a smoker all his life. He had quite a nasty area on his wound that was probably going to take quite a while to heal. I could just present him with the factors that I know about smoking, and encourage him probably to reduce that intake, that we all know.' (Clinician 18)

Providing information or advice on how to change
'I asked him...if it was time that he thought he could probably give up smoking, that this was... impairing his breathing...and I pointed out to him that I could probably help him, refer him to a quit smoking campaign and he said he would like to be able to stop smoking but he can't so...I just left it with him...and if he felt that he needed, he wanted to pursue it then I could point him in the right direction to do that. That's all I can do in that situation....' (Clinician 20)

Change support
'The client last week said she'll cut down on her drinking. She is pregnant...She only drinks six cans of bourbon and coke a day now, probably half what she'd normally, we try and build up a little helping network around them and try and sort out why they are acting like that, we need to help them change their living environments or think that there is help to do it' (Clinician 10)

them. For example, referring clients onward to more specialist service was a one off task requiring minimal skill and investment of time. In contrast, providing personalised support for lifestyle change required skills in behaviour change counselling and more time to engage clients in the change process that often occurred over a number of consultations. The choice of intervention strategies used largely reflected clinicians' intervention role expectations and intentions, as discussed in the previous section.

Weighing up the Benefits and Costs — Is it Worth it?

Finally, clinicians' appraisal of the overall benefits versus costs of their risk factor management practices acted to positively or negatively reinforce their commitment to addressing lifestyle issues (Figure 1). Some clinicians expressed uncertainty about whether addressing risk factors was a worthwhile component of their role because of their limited capacity for implementation (labelled role insufficiency), suggesting that perhaps this should be taken on by others. Other clinicians argued that the costs in terms of time and potential client resistance were not justified, given the limited perceived benefits in their client group. These clinicians expressed resentment that risk factor screening was a requirement of the service (labelled 'role resistance'), as illustrated in this quote:

'There have been no benefits [of the project] but extra work...At least half hour, if not an hour of extra work...Per client...with a negative result.' (Clinician 22)

In contrast, others endorsed risk factor management as a worthwhile practice due to the potential benefits of intervening and their capacity for implementation, resulting in professional satisfaction (labelled role verification), as summed up by this clinician:

'It hasn't been difficult to incorporate... I think in fact it's quite good to have some salient points to hit upon and it really hasn't made the assessment process that much unduly long...I think it's a very positive thing.it's what community health is all about.' (Clinician 4)

Discussion

The theoretical model presented in this study extends previous descriptive and cross-sectional studies by providing insight into the process by which clinician beliefs and attitudes shape the implementation of risk factor management in routine

PHC practice. Given the many competing demands facing PHC clinicians and their inability to address all preventive care needs [53,54], the findings suggest that clinicians rationalise their approach to managing lifestyle risk factors. This involves making judgements about the extent to which addressing lifestyle issues is considered a legitimate, doable, and worthwhile component of the role. The model suggests that implementation reflects both clinician beliefs about whether they should (commitment) and can (capacity) address lifestyle issues, and these beliefs are shaped by a range of patient, provider, and contextual factors. Beliefs about commitment and capacity, together with moral views about appropriate ways to intervene, all shape clinicians intentions about how they will intervene. This then provides a cognitive framework guiding their risk factor management practices. Finally, clinicians appraisal of the overall benefits and costs of address-ing lifestyle issues acts to positively or negatively reinforce their commitment to implementing these practices.

The model constructs are largely in line with previous quantitative and quali-tative studies suggesting that a combination of patient, contextual, and provider factors shape clinicians management of lifestyle risk factors. For example, previ-ous studies have found that higher risk patients [14,32,55-57], those perceived to be more motivated [58], and the least disadvantaged [55,59,60] are more likely to receive lifestyle intervention In line with our findings, contextual factors related to the service delivery environment have also been found to influence practices in previous studies including the length and number of consultations [55,61], pro-vider workload [62], and purpose of the visit [32,59,60]. Similarly, access to role support, such as training [18,55,63], decision support tools [32,36,55,63,64], collegial support [58,65], and client education materials [66-68] have all been associated with provision of lifestyle intervention.

Our findings offer fresh insights by suggesting that these patient and con-textual factors shape practice through their influence on providers' beliefs and attitudes. For example, clinicians are more committed to providing intervention to patients considered to be highly motivated because they perceive that these patients will be receptive, and the scope to make a difference is high. Similarly, access to role support and the service delivery context all influence perceptions about capacity. The model also highlights the synergistic relationship between commitment and capacity. Clinicians who perceive that they have the capacity to address lifestyle issues are more likely to believe that intervening will have a positive impact (scope to make a difference) reinforcing their commitment. Cli-nician commitment appears to be a prerequisite for capacity-building interven-tions to be effective. Finally, the model is unique in suggesting that beliefs about commitment and capacity, together with moral views about appropriate ways to intervene, all shape clinicians intentions about how they will intervene, which in turn determines the type of intervention strategies used.

At a theoretical level, the model has much in common with a model developed by Shaw and colleagues in the management of alcohol and other drugs (AOD) [69]. Shaw's model suggests that role perceptions, in particular role legitimacy (perceived boundaries of professional responsibility and right to intervene) and role adequacy (self-efficacy) form the foundation for health professionals motivation and satisfaction to respond to AOD issues [69,70]. Role support (help and advice from colleagues, supervisors, and other organisations), AOD education, and work experience were in turn thought to influence these role perceptions [69]. However, the current model extends Shaw's model in a number of ways. Firstly, it suggests that beliefs about outcomes, in particular beliefs about the 'scope to make a difference' and appraisal of benefits versus costs, are important in shaping commitment/motivation. It also expands the concept of role adequacy beyond self-efficacy to also include the extent which addressing lifestyle issues fits with current ways of working (self-delivery congruence). It suggests that role perceptions shape practice through intentions/expectations that also reflect philosophical views about appropriate ways to intervene.

The constructs identified in the model are largely in line with the main theoretical domains suggested by dominant psychological theories of motivation and action [71-76], including beliefs about capabilities, beliefs about consequences, and normative beliefs. Research suggests that these domains also apply to health professional practice, explaining on average 31% of 59% of the variance in clinician behaviour and intentions respectively [77]. These theories focus predominantly on individual cognitive factors shaping behaviour and do not explicitly include contextual/organisational factors and role beliefs. This is not surprising, given that these theories were developed to understand individual health behaviours rather than clinical practices. In contrast, the study model explicitly identifies the importance of the service delivery environment, as well as role beliefs—in particular beliefs about role congruence, role support, and intervention role expectations and intentions—in shaping risk factor management practices. As a result, the model provides new insights into theoretical constructs likely to be important in understanding the management of lifestyle risk factors that may apply more broadly to other clinician behaviours.

The study findings point to a number of possible leverage points for interventions to improve the lifestyle risk factor management practices of PHC clinicians. First, consideration should be given to tailoring the approach to lifestyle screening and intervention to suit the commitment and capacity of various healthcare providers. The findings suggest that it may be unrealistic to expect most providers to undertake all steps recommended in the widely endorsed 5A's approach to brief intervention [10,12]. A minimal approach to intervention would be to refer clients requiring intervention onwards to support services (arrange). This approach requires a minimal investment of time, and may best suited to clinicians for whom lifestyle issues are a peripheral component of their role and care is focused on treating a specific problem. The next level of intervention may be to provide brief advice regarding lifestyle recommendations followed by referral (advise and arrange). More intensive interventions, such as providing personalised support for lifestyle change tailored to the clients readiness to change with ongoing follow up and/or referral (advice, assist and arrange), is probably best suited to clinicians for whom lifestyle intervention is central to their role, the model of care is focused on early intervention/prevention, and they have specific knowledge and skills related to behaviour change interventions. At a system level, this tailored approach may be more realistic and facilitate uptake of practices and overall reach of lifestyle intervention to individuals. There is also evidence that minimal approaches (such as asking or brief advice) provided to individuals by more than one health professional can be effective in promoting behaviour change [78].

Second, improving practices is likely to require a range of professional development activities focusing on building positive clinician attitudes, skills, and self-efficacy. In particular, developing skills in behaviour change counselling, such as motivational interviewing approaches, is likely to be important in reducing client resistance and creating positive and effective interactions with clients compared to didactic approaches [79]. The findings suggest that shifting clinicians' views about the value and impact of lifestyle intervention is critical to enhancing commitment. This is likely to require a fundamental shift from a predominantly medical worldview, one that values high-tech interventions and dramatic outcomes, to a behavioural worldview that employs low-tech interventions (talking to people) to achieve small incremental changes in behaviour over time [80]. Given that professional values and norms are transmitted through early professional training, core competencies for lifestyle risk factor management should be integrated into undergraduate and postgraduate training for PHC providers. For existing providers, the inclusion of lifestyle risk factor management into continuing professional development activities will require a systematic and coordinated approach at the service/organisational level.

Third, at a service and organisational level, our findings highlight the importance of the model of service delivery and access to role support. In particular, service delivery models that involve case management, offer continuity of care, and focus on early intervention and prevention are likely to be more conducive to implementing risk factor management. This is likely to require policies to support service delivery redesign to improve the balance of preventive care and illness management. Embedding risk factor management activities into existing routines and work processes is also likely to be important along with wider organisational commitment required to sustain role supports, such as access to ongoing training, referral services, and decision support tools. Critical to achieving this will be a supportive policy environment for preventive care in PHC settings. These recommendations are consistent with the elements of the chronic care model that has recently been applied to health risk behaviours [81].

The study has a number of limitations. The model was developed in one PHC setting (state-funded community health services in NSW, Australia) based on the experiences of a limited number of participants and teams. Teams were selected to participate in the feasibility study based on an expression of interest, and hence may have been more interested and motivated to address lifestyle risk factors compared to other teams. Furthermore, participants who agreed to be interviewed may be more engaged in addressing lifestyle issues than participants who declined to be interviewed. However, a range of participants took part in the interviews, including those who felt positive and negative about the project. Furthermore, purposeful and theoretical sampling techniques ensured a wide range of participant were included in the analysis, including clinicians with both high and low levels of practice and a range of different types of practitioners (allied health professionals, registered and enrolled nurses, although all were female) from across the three teams. Insights were also gained from a number of other perspectives, including project officers involved in implementing the capacity building intervention, team managers, senior managers, and observations and reflections obtained by the researcher through prolonged engagement with the teams. Despite this, uncertainty remains about the extent to which model might apply to other PHC settings, such as general practice and other health professional behaviours. Further research is required to assess the usefulness of the model in other settings and contexts. In keeping with a constructivist approach, it is acknowledged that the model has been constructed based on the shared experiences between researchers and participants, and it aims to offer insights and understanding rather than verifiable knowledge.

Conclusion

The theoretical model presented in this paper suggests that clinician beliefs and attitudes shape the implementation of lifestyle risk factor management through the process of 'practice justification.' This involves justifying risk factor management practices as a legitimate, 'doable,' and worthwhile component of the role. The model offers new theoretical insights by suggested the importance of the service delivery environment and role beliefs in shaping practices in addition to individual cognitive factors suggested by psychological theories of motivation and action. Improving practices will not only require a range of professional development activities to build positive clinician attitudes and skills, but attention should be paid to creating models of service delivery conducive to preventive care and providing ongoing role support. Finally, consideration should be given to tailoring the approach to lifestyle screening and intervention to suit the commitment and capacity of various healthcare providers to maximise the reach of lifestyle screening and intervention at the population level. Further research is required to test the model, in particular its application in other settings, and to develop and test the effectiveness of strategies for improving the management of lifestyle risk factors in PHC.

Competing Interests

The authors declare that they have no competing interests.

Authors' Contributions

RL conducted the qualitative data analysis, contributed to study design and data collection, and wrote the first draft of the manuscript. LK and MH contributed to study design and data analysis. GPD and AW contributed to study design, while REM contributed to study design and data collection. All authors read, contributed to, and approved the final manuscript.

Acknowledgements

This study forms part of the PhD thesis undertaken by the first author (RL), supported by a scholarship from the National Health and Medical Research Council of Australia. The authors would like to acknowledge the Centre for Health Advancement, NSW Department of Health for funding the feasibility project which provided the data for this study along with the Community Health SNAP (smoking, nutrition, alcohol and physical activity) Project Team (Harris MF, Powell Davies G, Laws R, Williams A, Eames-Brown R, Amoroso C, Harper M, Greatz R, Gorrick P, Senuik S, Fuller S, Gilkes S, Angus L, Young C, Roe K, Jacobs S, Hughes J, Kehoe P). The findings presented represent the views of participants and do not necessarily represent the views of NSW Department of Health or Area Health Services. We would like to acknowledge services participating in the feasibility project, in particular the project officers who assisted with data collection, and to the providers who gave their time to participate.

References

1. WHO: The World Health Report: Reducing risks, promoting healthy lifestyle. Geneva. 2002.

2. Mokdad A, Marks J, Stroup D, Gerberding J: Actual causes of death in the United States. Journal of the American Medical Association 2004, 291:1238–1245.

3. Glasgow R, Goldstein M, Ockene J, Pronk N: Translating what we have learned into practice. Principles and hypotheses for interventions addressing multiple behaviors in primary care. Amercian Journal of Preventive Medicine 2004, 27:88–101.

4. Lancaster T, Stead L: Physician advice for smoking cessation (Review). Cochrane Database of Systematic Reviews 2004, CD000165.

5. Kaner E, Beyer F, Dickinson H, Pienaar E, Campbell F, Schlesinger C, Heather N, Saunders J, Burnand B: Effectiveness of brief alcohol interventions in primary care. Cochrane Database of Systematic Reviews 2007, (2):CD004148.

6. Goldstein M, Witlock E, DePue J: Multiple behavioural risk factor interventions in primary care. Amercian Journal of Preventive Medicine 2004, 27:61–79.

7. Lawton B, Rose S, Elley R, Dowell A, Fenton A, Moyes S: Exercise on prescription for women aged 40-74 recruited through primary care: two year randomised controlled trial. BMJ 2008, 337:a2509.

8. Counterweight Project Team: Evaluation of the Counterweight Programme for obesity management in primary care: A starting point for continuous improvement. British Journal of General Practice 2008, 58:548.

9. Whitlock EP, Polen MR, Green CA, Orleans T, Klein J: Clinical guidelines. Behavioral counseling interventions in primary care to reduce risky/harmful alcohol use by adults: a summary of the evidence for the U.S. Preventive Services Task Force. Annals of Internal Medicine 2004, 140:557.

10. Whitlock E, Orleans T, Pender N, Allan J: Evaluating primary care behavioural counseling interventions: An evidence-based approach. American Journal of Preventive Medicine 2002, 22:267–284.

11. Royal Australian College of General Practitioners: A population health guide to behavioural risk factors in general practice. Melbourne: RACGP; 2004.

12. US Preventive Services Taskforce: Treating Tobacco use and Dependence: 2008 Update. US Department of Health and Human Services; 2008.

13. Britt E, Miller G, Charles J, Henderson J, Bayram C, Harrison C, Valenti L, Fahridin S, Pan Y, O' Halloran J: General Practice Activity in Australia 2007–2008. Canberra: AIHW Cat No, GEP 22; 2008.

14. Ferketich AK, Khan Y, Wewers ME: Are physicians asking about tobacco use and assisting with cessation? Results from the 2001–2004 national ambulatory medical care survey (NAMCS). Preventive Medicine 2006, 43:472–476.

15. Heaton PC, Frede SM: Patients' need for more counseling on diet, exercise, and smoking cessation: Results from the National Ambulatory Medical Care Survey. Journal of the American Pharmacists Association 2006, 46:364–369.

16. Stange K, Flocke S, Goodwin M, Kelly R, Zyzanski S: Direct observation of rates of preventive service delivery in community family practice. Preventive Medicine 2000, 31:167–176.

17. Flocke S, Clark A, Schlessman K, Pomiecko G: Exercise, diet and weight loss advice in the family medicine outpatient setting. Family Medicine 2005, 37:415–421.

18. Schnoll RA, Rukstalis M, Wileyto EP, Shields AE: Smoking Cessation Treatment by Primary Care Physicians. An Update and Call for Training. American Journal of Preventive Medicine 2006, 31:233–239.

19. Ackermann R, Deyo R, LoGerfo J: Prompting primary providers to increase community exercise referrals for older adults: A randomized trial. J Am Geriatr Soc 2005, 53:283–289.

20. Aalto M, Pekuri P, Seppä K: Primary health care professionals' activity in intervening in patients' alcohol drinking during a 3-year brief intervention implementation project. Drug & Alcohol Dependence 2003, 69:9–14.

21. Andrews J, Tingen M, Waller J, Harper R: Provider feedback improves adherence with AHCPR smoking cessation guideline. Preventive Medicine 2001, 33:415–421.

22. Young J, D'Este C, Ward J: Improving family physicians' use of evidenc-based smoking cessation strategies: A cluster randomization trial. Preventive Medicine 2002, 35:572–583.

23. Douglas F, Teijlingen E, Torrance N, Fearn P, Kerr A, Meloni S: Promoting physical activity in primary care settings: health visitors' and practice nurses' views and experiences. Journal of Advanced Nursing 2006, 55:159–168.

24. Fuller T, Backett-Milburn K, Hopton J: Healthy eating: the views of general practitioners and patients in Scotland. American Journal of Clinical Nutrition 2003, 77:1043S-1047S.

25. Lawlor D, Keen S, Neal R: Can general practitioners influence the nation's health through a population approach to provision of lifestyle advice? Bristish Journal of General Practice 2000, 50:455–459.

26. Puffer S, Rashidan A: Practice nurses' intentions to use clinical guidelines. Journal of Advanced Nursing 2004, 47:500–509.

27. Edwards D, Freeman T, Gilbert A: Pharmacists' role in smoking cessation: an examination of current practice and barriers to service provision. International Journal of Pharmacy Practice 2006, 14:315–317.

28. Edwards D, Freeman T, Roche A: Dentists and dental hygienists' role in smoking cessation: An examination and comparison of current practices and barriers to service provision. Health Promotion Journal of Australia 2006, 17:145–151.

29. Sciamanna C, DePue J, Goldstein M, Park E, Gans K, Monroe A, Reiss P: Nutrition counseling in the promoting cancer prevention in primary care study. Preventive Medicine 2002, 35:437–446.

30. Brotons C, Ciurana R, Pineiro R, Kloppe P, Godycki-Cwirko M, Sammut M: Dietary advice in clinical practice: the view of general practitioners in Europe. American Journal of Clinical Nutrition 2003, 77(suppl):1048S–1051S.

31. Hutchings D, Cassidy P, Dallolio E, Pearson P, Heather N, Kaner E: Implementing screening and brief alcohol interventions in primary care: views from both sides of the consultation. Primary Health Care Research and Development 2006, 7:221–229.

32. Litaker D, Flocke S, Frolkis J, Stange K: Physicians attitudes and preventive care delivery: insights from the DOPC study. Preventive Medicine 2005, 40:556–563.

33. McIlvain H, Backer E, Crabtree B, Lacy N: Physician attitudes and the use of office-based activities for tobacco control. Family Medicine 2002, 34:114–119.

34. Kushner R: Barriers to providing nutrition counseling by physicians: a survey of primary care practitioners. Preventive Medicine 1995, 24:546–552.

35. Brotons C, Bjorkelund C, Bulc M, Ciurana R, Godycki-Cwirko M, Jurgova E, Kloppe P, Lionis C, Mierzecki A, Pineiro R, et al.: Prevention and health promotion in clinical practice: the views of general practitioners in Europe. Preventive Medicine 2005, 40:595–601.

36. Braun BL, Fowles JB, Solberg LI, Kind EA, Lando H, Pine D: Smoking-related attitudes and clinical practices of medical personnel in Minnesota. American Journal of Preventive Medicine 2004, 27:316–322.

37. Borrelli B, Hecht J, Papandonatos G, Emmons K, Tatewosian L, Abrams D: Smoking-cessation counseling in the home - Attitudes, beliefs, and behaviours of home healthcare nurses. Amercian Journal of Preventive Medicine 2001, 21:272–277.

38. Kviz F, Clark M, Prohaska T, Slezak J, Crittenden K, Freels S, Campbell F: Attitudes and practices for smoking cessation counselling by provider type and patient age. Preventive Medicine 1995, 24:201–212.

39. Pelkonen M, Kankkunen P: Nurses' competence in advising and supporting clients to cease smoking: a survey among Finnish nurses. Journal of Clinical Nursing 2001, 10:437–441.

40. Laws R, Kirby S, Powell Davies G, Williams A, Jayasinghe U, Amoroso C, Harris M: "Should I and can I?": A mixed methods study of clinician beliefs and attitudes in the management of lifestyle risk factors in primary health care. BMC Health Services Research 2008, 8:44.

41. Glaser B, Strauss A: The discovery of grounded theory. New York: Aldine Publishing Company; 1967.

42. Strauss A, Corbin J: Basics of Qualitative Research: Techniques and Procedures for Developing Grounded Theory. London: Sage; 1998.

43. Charmaz K: Constructing Grounded Theory: A Practical Guide Through Qualitative Analysis. London: Sage Publications; 2006.

44. Laws R, Powell Davies P, Williams A, Eames-Brown R, Amoroso C, Harris M: Community Health Risk Factor Management Research Project: Final Report. Sydney: Centre for Primary Health Care and Equity, UNSW; 2008.

45. Laws R, Williams A, Powell Davies G, Eames-Brown R, Amoroso C, Harris M: A square peg in a round hole? Approaches to incorporating lifestyle counselling into routine primary health care. Australian Journal of Primary Health 2008, 14:101–111.

46. NSW Health: Integrated Primary and Community Health Policy 2007–2012. Sydney: NSW Health; 2006.

47. Charmaz K: Coding in Grounded Theory Practice. In Constructing Grounded Theory: A Practical Guide Through Qualitative Analysis. Edited by: Silverman D. London: Sage Publications; 2006:42–71.

48. Charmaz K: Reconstructing theory in grounded theory studies. In Constructing Grounded Theory: A Practical Guide Through Qualitative Analysis. Edited by: Silverman D. London: Sage; 2006:123–150.

49. McGee G, Marland G, Atkinson J: Grounded theory research: literature reviewing and reflexivity. Journal of Advanced Nursing 2007, 60:334–343.

50. Charmaz K: Theoretical sampling, saturation, and sorting. In Constructing Grounded Theory: A Practical Guide Through Qualitative Analysis. Edited by: Silverman D. London: Sage Publications; 2006:42–71.

51. QSR International: NVivo (version 7). QSR International Pty Ltd; 2007.

52. Hall W, Callery P: Enhancing the rigor of grounded theory: Incorporating reflexivity and relationality. Qualitative Health Research 2001, 11:257–272.

53. Jaen CR, Stange KC, Nutting PA: Competing demands of primary care: A model for the delivery of clinical preventive services. Journal of Family Practice 1994, 38:166–171.

54. Yarnall K, Pollak K, Krause K, Michener J: Is there enough time for prevention? American Journal of Public Health 2003, 93:634–641.

55. Kaner EFS, Heather N, Brodie J, Lock CA, McAvoy BR: Patient and practitioner characteristics predict brief alcohol intervention in primary care. British Journal of General Practice 2001, 51:822–827.

56. Lock C, Kaner E: Implementation of a brief alcohol intervention by nurses in primary care: do non-clinical factors influence practice? Family Practice 2004, 21:270–275.

57. Burman ML, Kivlahan D, Buchbinder M, Broglio K, Zhou XH, Merrill JO, McDonell MB, Fihn SD, Bradley KA: Alcohol-related advice for veterans affairs primary care patients: Who gets it? Who gives it? Journal of Studies on Alcohol 2004, 65:621–630.

58. Sussman A, Williams R, Leverence R, Gloyd P, Crabtree B: The art and complexity of primary care clinicians' preventive counseling decisions: Obesity as a case study. Annals of Family Medicine 2006, 4:327–333.

59. Laws R, Jayasinghe U, Harris M, Williams A, Powell Davies G, Kemp L: Explaining the variation in the management of lifestyle risk factors in primary health care: A multilevel cross sectional study. BMC Public Health 2009, 9:165.

60. Ory M, Yuma P, Hurwicz M, Jarvis C, Barron K, Tai-Seale T, Tai-Seale M, Patel D, Hackethorn D, Bramson R, et al.: Prevalence and correlates of doctor-geriatric patient lifestyle discussions: Analysis of ADEPT videotapes. Preventive Medicine 2006, 43:494–497.

61. Berner MM, Harter M, Kriston L, Lohmann M, Ruf D, Lorenz G, Mundle G: Detection and management of alcohol use disorders in German primary care influenced by non-clinical factors. Alcohol and Alcoholism 2007, 42:308–316.

62. Pelletier-Fleury N, Le Vaillant M, Hebbrecht G, Boisnault P: Determinants of preventive services in general practice: A multilevel approach in cardiovascular domain and vaccination in France. Health Policy 2007, 81:218–227.

63. Amoroso C, Hobbs C, Harris M: General practice capacity for behavioural risk factor management: A SNAP-shot of a needs assesssment in Australia. Australian Journal of Primary Health 2005, 11:120–126.

64. Frank O, Litt J, Beilby J: Preventive activities during consultations in general practice: influences on performance. Australian family physician 2005, 34:508–512.

65. Sussman AL, Williams RL, Leverence R, Gloyd PW Jr, Crabtree BF: Self determination theory and preventive care delivery: A research involving outpatient settings network (RIOS Net) study. Journal of the American Board of Family Medicine 2008, 21:282–292.

66. Anis NA, Lee RE, Ellerbeck EF, Nazir N, Greiner KA, Ahluwalia JS: Direct observation of physician counseling on dietary habits and exercise: Patient, physician, and office correlates. Preventive Medicine 2004, 38:198–202.

67. Abatemarco D, Steinberg M, Denevo C: Midwives' knowledge, perceptions, beliefs and practice supports regarding tobacco dependent treatment. Journal of Midwifery & Womens Health 2007, 52(5):451–7.

68. Gottlieb NH, Guo JL, Blozis SA, Huang PP: Individual and contextual factors related to family practice residents' assessment and counseling for tobacco cessation. Journal of the American Board of Family Practice 2001, 14:343–351.

69. Shaw A, Cartwright A, Spratley T, Harwin J: Responding to Drinking Problems. London: Croom Helm; 1978.

70. Skinner N, Roche AM, Freeman T, Addy D: Responding to alcohol and other drug issues: The effect of role adequacy and role legitimacy on motivation and satisfaction. Drugs: Education, Prevention and Policy 2005, 12:449–463.

71. Ajzen I: The theory of planned behaviour. Organizational Behaviour and Human Decision Processes 1991, 50:179–211.

72. Bandura A: Self efficacy: towards a unifying theory of behavior change. Psychological Review 1977, 84:191–215.

73. Blackman D: Operant conditioning: an experimental analysis. London: Methuen; 1974.

74. De Vries H, Dijkstra M, Kuhlman P: Self-efficacy: The third factor besides attitude and subjective norm as a predictor of behavioural intentions. Health Education Research 1988, 3:273–282.

75. Prochaska J, Velicer W: The transtheoretical model of health behaviour change. American Journal of Health Promotion 1997, 12:38–48.

76. Valois P, Desharnais R, Godin G: A comparison of the Fishbein and Ajzen and the Triandis attitudinal models for the prediction of exercise intention and behavior. Journal of Behavioral Medicine 1988, 11:459–472.

77. Godin G, Belanger-Gravel A, Eccles M, Grimshaw J: Healthcare professionals intentions and behaviours: A systematic review of studies based on social cognitive theories. Implementation Science 2008, 3:36.

78. Lawrence C, Folders S, NL A, Bluhm J, Bland P, Davern M, Schillo B, Ahluwalia J, Manley M: The impact of smoking-cessation intervention by multiple health professionals. American Journal of Preventive Medicine 2008, 34:54–60.

79. Rubak S, Sandbaek A, Lauritzen T, Christensen B: Motivational interviewing: a systematici review and meta-analysis. Bristish Journal of General Practice 2005, April:305–310.

80. Botelho R: Motivational Practice: Promoting healthy habits and self-care of chronic diseases. 2nd edition. New York: MHH Publications; 2004.

81. Hung D, Rundall T, Tallia A, Cohen D, Halpin H, Crabtree B: Rethinking prevention in primary care: Applying the chronic care model to address health risk behaviors. The Milbank Quarterly 2007, 85:69–91.

Prevalence and Risk Factors for Stunting and Severe Stunting Among Under-Fives in North Maluku Province of Indonesia

Ramli, Kingsley E. Agho, Kerry J. Inder, Steven J. Bowe, Jennifer Jacobs and Michael J. Dibley

ABSTRACT

Background

Adequate nutrition is needed to ensure optimum growth and development of infants and young children. Understanding of the risk factors for stunting and severe stunting among children aged less than five years in North Maluku province is important to guide Indonesian government public health planners to develop nutrition programs and interventions in a post conflict area.

The purpose of the current study was to assess the prevalence of and the risk factors associated with stunting and severe stunting among children aged less than five years in North Maluku province of Indonesia.

Methods

The health and nutritional status of children aged less than five years was assessed in North Maluku province of Indonesia in 2004 using a cross-sectional multi-stage survey conducted on 750 households from each of the four island groups in North Maluku province. A total of 2168 children aged 0-59 months were used in the analysis.

Results

Prevalence of stunting and severe stunting were 29% (95%CI: 26.0-32.2) and 14.1% (95%CI: 11.7-17.0) for children aged 0-23 months and 38.4% (95%CI: 35.9-41.0) and 18.4% (95%CI: 16.1-20.9) for children aged 0-59 months, respectively. After controlling for potential confounders, multivariate analysis revealed that the risk factors for stunted children were child's age in months, male sex and number of family meals per day (≤2 times), for children aged 0-23 months, and income (poorest and middle-class family), child's age in months and male sex for children aged 0-59 months. The risk factors for severe stunting in children aged 0-23 months were income (poorest family), male sex and child's age in months and for children aged 0-59 months were income (poorest family), father's occupation (not working), male sex and child's age in months.

Conclusion

Programmes aimed at improving stunting in North Maluku province of Indonesia should focus on children under two years of age, of male sex and from families of low socioeconomic status.

Background

The optimal growth and development of infants and young children are fundamental for their future [1]. Stunting, a deficit in height or length relative to a child's age is a major health problem in South Asia where half of children aged less than five years are stunted [2]. In Indonesia, 37% of children aged less than five years are stunted [3]. Promoting better eating habits in an effort to improve nutrition is one of the most challenging tasks in Indonesia as malnutrition remains one of the most important public health problems facing almost every district [4].

In Indonesia, like many developing countries, the most common nutritional problems in infancy and early childhood are stunting, wasting; iron-deficiency anaemia, poverty and low birth weight [5,6]. Malnutrition during the first 2 years of life can lead to mortality and morbidity in childhood [7,8] and is one of the most preventable risk factors for mortality [9].

Past studies have also shown that lower intelligence quotient (IQ), mother's height, male sex, mother and father level of education, poverty, socioeconomic status, residence, child care behaviour (inadequate complimentary feeding and breastfeeding), cultural beliefs, access to health care and environmental ecosystems [10,11] are factors associated with stunting in children aged less than five years.

Despite the persistently high prevalence of stunted children in Indonesia, there is a lack of information about the prevalence and risk factors associated with stunted and severely stunted children in the North Maluku province of Indonesia using the new Growth reference from the World Health Organisation [12]. This province is an area in Indonesia that in 2004 had recently emerged from a period of prolonged civil conflict. This paper assesses the prevalence and risk factors associated with stunting and severe stunting in children aged 0-59 months old in North Maluku province of Indonesia.

Methods

Study Location

The study covered all areas in the North Maluku province of Indonesia (see Figure 1) [13] with a total population of about 920,000 people in 2006 [14,15], divided

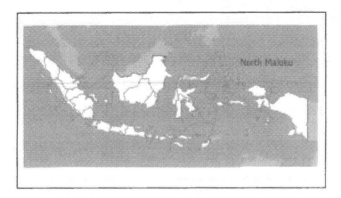

Figure 1. Map of Indonesia showing North Maluku Province. (Source: Wikipedia, 2008)

into four island groups The first island group consists of the districts of Ternate and Tidore with a total population of about 241,000 people. The second island group consists of the districts of Central Halmahera and East Halmahera with a total population of about 95,000 people. The third island group consists of the districts of West Halmahera and North Halmahera with a total population of about 276,000 people. The forth island group consists of the districts of South Halmahera and Sula-Isles with a total population of about 308,000 people [15].

Study Design

A cross-sectional survey was conducted in 2004 on 3000 households from the four island groups in North Maluku province. A multistage cluster sampling technique was used for selecting the study sample in which North Maluku province was grouped into four island groups with eight districts in total.

Selection of Subjects

The four island groups within North Maluku province were used to select the study areas. In the first stage, two districts were randomly selected based on probability proportional to size from each island group [15]. In the second stage, subdistricts (referred to as clusters) were randomly selected from each district. In the third stage, Puskesmas (Community Public Health Services) were selected randomly from each subdistrict and finally, the villages were randomly selected from Puskesmas. Household selection in each cluster was randomly taken by using the sampling frame of every 10th household with the nearest household from the village health service (Pustu) as the starting point. Comprehensive details of the study districts and selection criteria have been reported elsewhere [16]. In total, 50 households were selected in each cluster and 15 clusters in each island group, yielding a total of 750 households from each island group.

Stunting (Height-for-Age)

The nutritional status of children less than five years of age was measured anthropometrically. Length was measured for children aged less than two years old and height for those two years of age and older. Length was measured using a wooden stadiometer to the nearest 0.1 cm and height was measured using Microtoice tape to the nearest 0.1 cm [17]. The height-for-age measurement status was expressed in Standard Deviation (SD) units (Z-score) from the median of the reference population. Children with a measurement of <-2 SD units from the median of the reference population were considered short for their age (stunted) and children with measurement of <-3SD units from the median of the reference population were considered to be severely stunted.

Socioeconomic Factors

A structured household questionnaire was used for collecting information about the following family level factors: region (urban and rural), district (total of eight), father's level of education (completed elementary school [6 years of schooling], completed middle school [9 years of schooling] and completed high school [12 years of schooling]; mother's level of education; parental education (both with higher education, father with high education, mother with high education, neither with high education); father's occupation, mother's occupation; parental occupation; household wealth index (calculated from household's ownership of consumer items including refrigerators, VCR players, satellite dishes, televisions, lounges, boats, cars, motorcycles; flooring material; type of drinking-water source; toilet facilities; and other characteristics that are related to wealth status—categorised into poorest, middle and least poor); number of household members in the family and number of family meals per day. In addition, the following child level factors were collected: child's age in months; gender; provision of nutritional status information during pregnancy and number of antenatal visits. The questionnaires were administered after a signed informed consent was collected. The questionnaires were checked daily for accuracy, consistency, and completeness by field supervisors.

Ethical Permission

This study had the approval of the Ministry of Health in Indonesia and the protocol for secondary data analysis was approved by Human Ethics Research Committee, University of Newcastle–Australia.

Statistical Analyses

Data were entered into a computerised database and cleaned using the data entry program EPIINFO [18]. Data on nutritional status were analysed based on the new growth reference from the World Health Organisation. The household ownership of consumer items described above were used in constructing the wealth index scores by using a method similar to that described by Filmer and Pritchett [19] and were divided into three categories. The bottom 40% of the households was referred to as the poorest households, the next 40% as the middle-class households, and the top 20% as the least poor households. The analysis for stunted and severely stunted children was categorised into two groups (1) children aged 0-23 months, and (2) children aged 0-59 months.

To determine the level of stunting and severe stunting the dependent variable was expressed as a dichotomous variable: category 0 if not stunted (≥-2SD) or severely stunted (≥-3SD) and category 1 if stunted (<-2SD) or severely stunted (<-3SD).

Firstly, univariate binary logistic regression analysis was performed to examine the association between stunted and severely stunted children aged 0-23 months and stunted children 0-59 months. Secondly, the factors associated with stunting and severe stunting were examined in a multiple logistic regression model. A stepwise backward elimination approach was applied. At the start variables were selected for inclusion in the model if their univariate analysis p-value was <= 0.25. Only variables which were statistically associated with stunted and severely stunted children (p < 0.05) remained in the final model. Unadjusted and adjusted odds ratios from a logistic model are presented with 95% confidence intervals. The 'SVY' commands from Stata version 9.2 (Stata Corp) were used for data analysis to adjust for the cluster sampling design and appropriate sampling weights.

Results

Univariate Analyses

Table 1 presents the prevalence of stunting and severe stunting in children aged 0-23 months and 0-59 months, respectively. Table 1 reveals that the child's age and gender were significantly associated with stunting in children aged 0-23 months while the mother's education, the household wealth index, the child's age and gender were significantly associated with severe stunting in children aged 0-23 months.

For children aged 0-59 months, parental education (only mothers with higher education and neither parent with higher education), household wealth index (poorest), region (urban), child's age and males were statistically significantly associated with stunting. While, father's occupation (not working), parental education (neither parent with higher education), household wealth index (poorest), child's age (24-29 months old) and males reported higher prevalence of severe stunting. The overall prevalence of being stunted and severely stunted was 29%

and 14.1% for children aged 0-23 months and 38.4% and 18.4% for children aged 0-59 months (see Table 1), respectively.

Table 1. Prevalence of stunting and severe stunting in children aged 0-23 months and 0-59 months

Characteristic	n	Stunted children 0-23 Months % (95% CI)	Severely stunted children 0-23 Months % (95% CI)	Stunted children 0-59 Months % (95% CI)	Severely stunted children 0-59 Months % (95% CI)
Region					
Rural	536	27.1 (21.0-34.3)	13.8 (9.5-19.5)	33.4 (28.6-38.6)ᵃ	15.9 (12.6-20.1)
Urban	1632	29.6 (26.2-33.2)	14.2 (11.4-17.6)	40.0 (37.2-42.9)	19.2 (16.4-22.3)
District					
Ternate	387	30.9 (23.8-39.1)	14.2 (9.9-19.9)	36.4 (31.0-42.2)	16.0 (12.3-20.6)
Tidore	149	19.1 (10.4-32.5)	12.8 (4.9-29.2)	25.5 (18.9-33.4)	15.4 (8.6-26)
Central Halmahera	280	32.3 (25.7-39.7)	15.5 (11.3-21)	42.9 (36.5-49.5)	23.2 (18.1-29.3)
East Halmahera	280	37.0 (28.3-46.7)	21.9 (13.2-34.3)	42.5 (33.2-52.4)	23.6 (14.6-35.8)
West Halmahera	178	23.7 (15.5-33.8)	6.2 (3.1-12.0)	38.2 (33.2-43.5)	16.3 (12.5-21)
North Halmahera	279	31.1 (24.5-38.6)	13.3 (9.1-19.1)	41.6 (34.5-49.0)	17.2 (12.6-23.0)
South Halmahera	446	24.7 (19.7-30.5)	12.4 (8.2-18.2)	38.6 (35.2-42.1)	19.3 (15.1-24.3)
Sula-Isles	169	27.4 (20.7-35.3)	10.7 (6.7-16.8)	34.3 (29.6-39.4)	11.2 (9-14.0)
Household factors					
Father's education					
Completed Elementary School (aged 7-12)	832	32.0 (27.2-37.2)	16.2 (12.6-20.7)	41.8 (37.8-45.9)ᵃ	20.9 (17.6-24.7)ᵃ
Completed Middle School (aged 13-15)	557	28.4 (23.4-34.0)	13.2 (9.7-17.6)	39.3 (35.2-43.6)	19.0 (15.5-23.2)
Completed High School (aged 16-18)	779	26.3 (22.0-31.2)	12.6 (9.2-17)	34.0 (30.3-37.9)	15.2 (12.1-18.8)
Mother's education					
Completed Elementary School (aged 7-12)	1163	30.8 (27.0-34.9)	16.7 (13.6-20.30)ᵃ	41.6 (38.4-44.9)ᵃᵃ	20.9 (18.0-24.3)ᵃᵃ
Completed Middle School (aged 13-15)	514	29.7 (24.5-35.5)	11.5 (8-16.4)	37.7 (32.8-43)	15.8 (12.4-19.8)
Completed High School (aged 16-18)	491	24.5 (19.6-30.2)	11.2 (7.7-16.1)	31.4 (27.0-36.0)	14.9 (11.7-18.8)
Parental Education					
Both with high education	871	26.2 (22.1-30.7)	11.3 (8.4-14.9)	33.6 (30-37.5)ᵃᵃ	15.4 (12.7-18.6)ᵃ
Father with high education	134	33.3 (22.7-46.0)	12.0 (5.6-23.7)	41.0 (32.6-50.1)	14.9 (9.4-22.9)
Mother with high education	465	29.2 (23.5-35.7)	16.1 (11.8-21.5)	41.1 (37.1-45.1)	19.4 (15.6-23.8)
Neither with high education	698	31.7 (26.9-37.1)	17.1 (13.1-21.9)	41.9 (37.7-46.4)	22.1 (18.4-26.2)
Provided with nutritional information during pregnancy (n = 2110)					
No	729	29.9 (25.4-35.0)	16.0 (12.8-19.9)	42.0 (37.7-46.4)	21.1 (17.6-25.1)
Yes	1381	28.2 (24.9-31.9)	12.9 (10.1-16.3)	36.6 (33.7-39.7)	17.0 (14.3-20.1)
Father's occupation					
Any Labour	1396	29.3 (25.6-33.3)	15.1 (12.2-18.7)	40.0 (36.9-43.2)	19.3 (16.7-22.3)ᵃᵃ
Fisher man	355	27.9 (22.6-33.8)	12.5 (8.5-18)	35.5 (30.8-40.5)	18.0 (14.5-22.1)
No work	60	35.9 (21.7-53.1)	20.5 (10.4-36.5)	45.0 (34.0-56.5)	28.3 (18.5-40.5)
Government private officer	357	27.7 (20.9-35.7)	10.8 (6.9-16.5)	33.9 (29-39.2)	13.2 (9.8-17.5)
Mother's occupation					
Any Labour	694	29.9 (25.3-34.9)	14.4 (11.4-18.2)	39.5 (35.4-43.8)	18.4 (15.4-22)
Fisher women	96	30.6 (20.6-42.8)	18.4 (9.8-31.8)	38.5 (30.8-46.9)	22.9 (15.5-32.5)
No work	1302	28.4 (24.4-32.7)	13.6 (10.4-17.6)	37.9 (34.5-41.4)	18.1 (15.2-21.5)
Government private officer	76	29.2 (17.7-44.0)	14.6 (8.0-25.0)	36.8 (27.6-47.2)	15.8 (10.2-23.7)
Parental employment					
Both working	838	29.8 (25.8-34.2)	14.9 (11.9-18.6)	38.7 (35.3-42.2)	18.1 (15.4-21.3)
Father only working	1270	28.1 (24.1-32.4)	13.2 (10.0-17.2)	37.9 (34.5-41.4)	18.0 (15.1-21.4)
Mother only working	28	33.3 (14.2-60.2)	13.3 (2.8-44.8)	53.6 (38.5-68.1)	35.7 (19.7-56.8)
Neither working	32	37.5 (21.1-57.4)	25.0 (11.7-45.7)	37.5 (23.1-54.6)	21.9 (11.4-38.0)
Household wealth Index					
Poorest	867	32.9 (27.6-38.7)	17.4 (13.4-22.4)ᵃᵃ	43.1 (39.0-44.0)ᵃᵃ	21.1 (17.4-25.4)ᵃᵃ
Middle	867	28.2 (24.3-32.5)	13.9 (11.0-17.6)	37.4 (33.9-40.9)	18.9 (16.3-21.9)
Least Poor	434	23.4 (18.1-29.7)	8.3 (5.1-13.4)	30.9 (26.4-35.8)	11.8 (8.7-15.7)

Table 1. *(Continued)*

Household member					
≤5 members	1359	31.3 (27.6-35.4)	14.8 (11.6-18.7)	39.3 (36.2-42.5)	18.5 (15.7-21.7)
6-12 members	809	24.9 (21-29.4)	12.9 (10.1-16.4)	36.8 (33.0-40.8)	18.1 (15.2-21.3)
Number of family Meals per day					
2 Times	675	24.9 (20.5-29.8)	13.1 (9.9-17.1)	36.3 (32.2-40.6)	17.6 (14.3-21.4)
>2 Times	1493	30.8 (27.0-34.8)	14.5 (11.5-18.2)	39.3 (36.22-42.5)	18.7 (16-21.8)
Child level factors					
Child's age in category					
0-5	264	12.8 (9.1-17.7)***	7.9 (5.6-11.0)***	12.8 (9.1-17.7)***	7.9 (5.6-11.0)***
6-11	365	24.1 (19.3-29.7)	10.1 (5.6-11.0)***	24.1 (19.3-29.7)	10.1 (6.6-15.2)
12-17	318	33.9 (28.6-39.7)	15.7 (5.6-11.0)***	33.9 (28.6-39.7)	15.7 (11.7-20.8)
18-23	234	48.3 (40.3-56.3)	25.2 (5.6-11.0)***	48.3 (40.3-56.3)	25.2 (18.8-32.9)
24-29	238			48.7 (42.2-55.4)	26.5 (20.4-33.6)
30-35	169			51.5 (42.9-60.0)	23.1 (16.8-30.8)
36-41	215			52.6 (45.9-59.1)	25.6 (20.5-31.5)
42-47	147			47.6 (37.7-57.7)	19.7 (13-28.8)
48-53	135			49.6 (39.9-59.4)	23.7 (16.8-32.3)
54-59	81			44.4 (34.6-54.8)	16.1 (8.7-27.6)
Gender					
Male	1115	32.2 (28.0-36.7)*	16.7 (13.2-21.1)	41.4 (37.9-44.8)*	20.5 (17.4-24.1)**
Female	1053	25.6 (21.6-30.1)	11.3 (8.9-14.3)	35.2 (31.8-38.8)	16.1 (13.7-18.7)
Antenatal visit (n = 2118)					
No	91	39.6 (27.2-53.4)	16.7 (7.6-32.7)	44.0 (35.1-53.2)	23.1 (15.4-33.1)
Yes	2019	28.4 (25.4-31.5)	16.7 (11.5-16.6)	38.2 (35.6-40.9)	18.2 (15.9-20.8)
Overall		29.0 (26.0-32.2)	16.7 (11.7-17.8)	38.4 (35.9-41.0)	18.4 (16.1-20.9)

*P < 0.05; **P < 0.01; ***P < 0.001; X²-test was applied to test statistical significance
Weighted total was 2168 otherwise stated within brackets

Multivariate Analyses

Tables 2 and 3 shows the unadjusted and adjusted odds ratios for the association between stunted and severely stunted children and socioeconomic characteristics of children aged 0-23 months children aged 0-59 months.

Risk Factors for Stunting

The odds for stunted children aged 0-23 months was 26 percent lower in families that provided at least three meals per day. Increased child age in months was statistically associated with stunting in children aged 0-23 months (adjusted OR (AOR) = 1.11, 95%CI: 1.08 - 1.14; p < 0.001) and girls had reduced odds of being stunted compared to boys (AOR = 0.67, 95%CI: 0.50 - 0.89; p = 0.006).

Children aged 0-59 months from families in the least poor or middle household wealth index categories had reduced odds of being stunted compared to those from the poorest families. Increased age of the child was statistically associated with stunting in children aged 0-59 months (AOR = 1.03, 95%CI: 1.02 - 1.04; p < 0.001). Girls aged 0-59 months had statistically significantly reduced odds of being stunted compared to boys aged 0-59 months (AOR = 0.72, 95% CI: 0.58 - 0.90; p = 0.005).

Table 2. Risk factors for stunting in children aged 0-23 months and 0-59 months

Characteristic	Stunted children 0-23 Months		Stunted children 0-59 Months	
	Unadjusted OR (95% CI)	Adjusted OR (AOR) (95% CI)	Unadjusted OR (95% CI)	AOR (95% CI)
Region				
Rural	1.00		1.00	
Urban	1.13 (0.77-1.65)		1.33 (1.03-1.71)	
District				
Ternate	1.00		1.00	
Tidore	0.53 (0.24-1.17)		0.60 (0.38-0.94)	
Central Halmahera	1.06 (0.66-1.72)		1.31 (0.91-1.88)	
East Halmahera	1.31 (0.76-2.24)		1.29 (0.81-2.06)	
West Halmahera	0.69 (0.38-1.28)		1.08 (0.78-1.50)	
North Halmahera	1.01 (0.62-1.64)		1.24 (0.84-1.83)	
South Halmahera	0.73 (0.46-1.16)		1.10 (0.82-1.45)	
Sula-Isles	0.84 (0.50-1.41)		0.91 (0.66-1.26)	
Household factors Father's education				
Completed Elementary School (aged 7-12)	1.00		1.00	
Completed Middle School (aged 13-15)	0.84 (0.59-1.21)		0.90 (0.71-1.14)	
Completed High School (aged 16-18)	0.76 (0.55-1.05)		0.70 (0.57-0.91)	
Mother's Education				
Completed Elementary School (aged 7-12)	1.00		1.00	
Completed Middle School (aged 13-15)	0.95 (0.69-1.31)		0.85 (0.66-1.09)	
Completed High School (aged 16-18)	0.73 (0.55-0.97)		0.64 (0.50-0.83)	
Parental Education				
Both with high education	1.00		1.00	
Father with high education	1.41 (0.78-2.54)		1.37 (0.92-2.04)	
Mother with high education	1.17 (0.83-1.64)		1.38 (1.09-1.73)	
Neither with high education	1.31 (0.96-1.80)		1.43 (1.11-1.84)	
Provided with nutritional information during pregnancy				
No	1.00		1.00	
Yes	0.78 (0.56-1.07)		0.77 (0.57-1.03)	
Father's occupation				
Any Labour	1.00		1.00	
Fisher man	0.93 (0.66-1.32)		0.83 (0.65-1.06)	
No work	1.35 (0.65-2.80)		1.23 (0.77-1.96)	
Government private officer	0.92 (0.62-1.38)		0.77 (0.60-0.99)	
Mother's occupation				
Any Labour	1.00		1.00	
Fisherwoman	1.03 (0.58-1.83)		0.96 (0.65-1.42)	
No work	0.93 (0.69-1.25)		0.93 (0.75-1.17)	
Government private officer	0.96 (0.49-1.92)		0.89 (0.55-1.46)	
Parental employment				
Both working	1.00		1.00	
Father only working	0.92 (0.70-1.21)		0.97 (0.79-1.18)	
Mother only working	1.18 (0.39-3.59)		1.83 (0.97-3.47)	
Neither working	1.41 (0.62-3.20)		0.95 (0.47-1.92)	
Household wealth index				
Poorest	1.00		1.00	1.00
Middle	0.80 (0.58-1.10)		0.87 (0.65-1.16)	0.78 (0.63-0.98)
Least Poor	0.62 (0.41-0.94)		0.50 (0.33-0.75)	0.62 (0.45-0.85)
Household member				
≤5 members	1.00		1.00	
6-12 members	0.73 (0.56-0.95)		0.90 (0.73-1.11)	

Table 3. Risk factors for severe stunting in children aged 0-23 months and 0-59 months

Risk Factors for Severe Stunting

As shown in Table 3 the AOR indicated that children aged 0-23 months from families in the least poor or middle household wealth index categories had reduced odds of being severely stunted compared to those from the poorest families. Increasing age of the child was significantly associated with severe stunting (AOR = 1.08, 95%CI: 1.05 - 1.12; p < 0.001). Boys aged 0-23 months had increased odds of being severely stunted compared with girls aged 0-23 months (AOR = 0.58, 95%CI: 0.42 - 0.81; p = 0.002).

Children aged 0-59 months from least poor families had reduced odds of being severely stunted (AOR = 0.52, 95%CI: 0.33 - 0.83; p = 0.005) compared with those from middle and poorest families. Increasing age of the child was significantly associated with severe stunting in children aged 0-59 months (AOR = 1.02, 95%CI: 1.01 - 1.03; p < 0.001). Boys aged 0-59 months had increased odds of being severely stunted compared to girls aged 0-59 months (AOR = 0.72, 95%CI: 0.58 - 0.90; p = 0.005).

Discussion

This paper presents data on the prevalence and risk factors associated with stunting and severe stunting in children in North Maluku province of Indonesia. This is the first study to assess the prevalence and factors associated with stunting and severe stunting in children aged less than five years in North Maluku province of Indonesia.

The prevalence of stunting in children in this population was high with 29% of the children aged 0-23 months and 38.4% of the children aged 0-59 months being stunted while, 14.1% and 18.4%, respectively were severely stunted. This level of stunting in North Maluku was higher than the national level among children aged 0-59 months of 28.6% reported in 2004 [20].

The prevalence of stunting and severe stunting was higher in children aged 24-59 months (50% vs 24%, respectively) than those children aged 0-23 months. This findings is similar to the results from Bangladesh, India and Pakistan [21-23] where children aged 24-59 months were found to be at a greater risk of being stunted. This suggests that for children aged 24-59 months stunting is not likely to be reversible [24]. Our findings support the previous assertions that the prevalence of stunting remains constant after 2 or 3 years of life [25].

Comparing the present study with children in four regions (Africa; Asia, Latin and South America), the prevalence of stunting in children 0-59 months in North Maluku province of Indonesia was slightly lower than that of Africa (40.1%), higher in Asia (31.3%), Latin America (16.1%) and South America (13.8%)[8]. These differences in prevalence likely result from a combination of factors like environmental, cultural differences and prolonged civil conflicts or war.

The results indicate that the gender of the child is a strong predictor of stunting and severe stunting in children aged 0-23 months and children aged 0-59 months. Girls had lower odds of becoming stunted or severely stunted compared to boys which supports the findings of other studies [26-28]. During infancy and childhood, girls were less likely to become stunted and severely stunted than boys and infant girls survive in greater numbers than infant boys in most developing countries including Indonesia [26-28].

Multivariate analyses identified that the following factors were statistically significantly associated with stunting and severe stunting for all of the three age categories after controlling for potential confounders: (a) number of family meals per day (2 meals per day); (b) child's age in months; (c) male sex; (d) household wealth (poorest); (e) parental employment (not working) and (f) district (Central and South Halmahera). These findings support similar studies indicating that mother's education, household wealth, gender, age and employment were significantly associated with stunted children [28-30].

This study suggests that stunting in children in North Maluku may be reduced by improving mother's education, mother's nutritional information and reducing poverty. This study highlights the need to provide special attention to reducing stunting in the Central and South Halmahera districts. Interventions for improving the provision of education for girls are required as lack of education appears to be a major risk factor for stunting in children in North Maluku province.

A number of important limitations due to the nature of the data used (secondary data analysis) needs to be considered. Firstly, there is no dietary intake data to support our findings. Secondly, this study used only two main characteristics (family and child level factors). Finally, the design is cross-sectional and reports only a "snapshot" of the frequency of stunting and severe stunting. Hence no strong conclusions can be made as to the possible causes of stunting and severe stunting.

Despite these limitations, the findings from this study contribute to our understanding of the factors associated with stunting and severe stunting in children aged 0-59 months in North Maluku, Indonesia. The findings in this study will assist the local government in North Maluku develop an appropriate intervention for the children aged 0-59 months and their parents. It may also help the Indonesian government public health planners develop a national nutrition program and interventions targeting young children especially in post conflict areas in eastern Indonesia. However, further research is required to understand the dietary and other determinants (including environmental risk factors) of stunting in North Maluku province of Indonesia.

Conclusion

Childhood malnutrition remains a major public health problem in Indonesia. Results from this cross-sectional study showed that child's age in months, low socioeconomic status and gender (being a male child) were significant risk factors for stunting and severe stunting in North Maluku province of Indonesia. These results highlight the need for early intervention programmes aimed at reducing undernutrition in children, especially in the first two years of life.

Competing Interests

The authors declare that they have no competing interests.

Authors' Contributions

R and MJD designed the study. R, KA and SJB carried out the statistical analysis. R and KA wrote the manuscript. All authors made contributions to the interpretation of results and revised the manuscript for important intellectual content. All authors read and approved the final version of the manuscript.

Acknowledgements

This analysis is a part of the first author's thesis to fulfill the requirement for a PhD in Medicine at Newcastle University and we would like to thank Professor Cate D'Este, Director of the Centre for Clinical Epidemiology and Biostatistics, University of Newcastle for her excellent cooperation.

References

1. Sguassero Y, de Onis M, Carroli G: Community-based supplementary feeding for promoting the growth of young children in developing countries (Review). [http://www.cochrane.org/reviews/en/ab005039.html] (assessed 12 Dec., 2008)

2. Bhutta ZA: Why has so little changed in maternal and child health in south Asia? Br Med J 2000, 321:809–812.

3. Atmarita TSF: A summary of the current nutrition situation in Indonesia. Paper presented at Capacity and Leadership Development in Nutritional Sciences, Seoul Korea September 4 – 6, 2008

4. Atmarita TSF: Nutrition problem in Indonesia. Paper presented at An Integrated International Seminar and Workshop on Lifestyle - Related Diseases, Gajah Mada University, 19 – 20 March, 2005

5. Black RE, Morris SS, Bryce J: Where and why are 10 million children dying every year? Lancet 2003, 361(9376):2226–34.

6. Arifeen S, Black RE, Antelman G, Baqui A, Caulfield L, Becker S: Exclusive breastfeeding reduces acute respiratory infection and diarrhea deaths among infants in Dhaka slums. Pediatrics 2001, 108(4):E67.

7. Pelletier D, Frongillo EA: Changes in Child Survival Are Strongly Associated with Changes in Malnutrition in Developing Countries. Journal of Nutrition 2003, 133:107–119.

8. Black R, for the Maternal and Child Undernutrition Study Group, et al.: Maternal and child undernutrition: Global and regional exposures and health consequences. Lancet 2008, 371:243–360.

9. Penny M, Creed-Kanashiro HM, Robert RC, Narro MR, Caulfield LE, Black RE: Effectiveness of an educational intervention delivered through the health services to improve nutrition in young children: a cluster randomized controlled trial. Lancet 2005, 365:1863–1872.

10. Ayaya SO, Esamai FO, Rotich J, Olwambula AR: Socio-economic factors predisposing under five-year-old children to severe protein energy malnutrition at the Moi Teaching and Referral Hospital, Eldoret, Kenya. East African Medical Journal 2004, 81(8):415–421.

11. Hautvast JL, et al.: Severe linear growth retardation in rural Zambian children: the influence of biological variables. Am J Clin Nutr 2000, 71(2):550–559.

12. World Health Organization: World Health Organization releases new Child Growth Standards. In "Standards confirm that all children worldwide have the potential to grow the same." WHO; Geneva; 2006.

13. Maluku-Utara: Pemerintahan Maluku Utara. [http://en.wikipedia.org/wiki/North_Maluku] 2008. (assessed on 26 June 2008)

14. Razak TA: Assessment of Education, Health, Nutrition, Water and Sanitation and Protection Needs of Children in North Maluku. Hasanuddin University Press, Makassar, Indonesia; 2004.

15. Badan Pusat Statistik: Statistic Indonesia. [http://www.bps.go.id/] (assessed on 12 Dec., 2008)

16. Ramli: Nutrition Status and Risk Factors for Malnutrition in Young Children in North Maluku province of Indonesia. PhD thesis. University of Newcastle, Australia; 2009.

17. World Health Organization: Physical Status: The use and interpretation of anthropometry, in WHO Technical Report Series No. 854. World Health Organ Tech Rep Ser 1995, 854:1–452.

18. Dean A, et al.: Epi-Info, version 5. Atlanta: Centers for Disease Control; 2000.

19. Filmer D, Pritchett LH: Estimating wealth effects without expenditure data - or tears: an application to educational enrolments in states of India. Demography 2001, 38:115–32.

20. World Health Organization: Global database on child and malnutrition. [http://www.who.int/nutgrowthdb/database/countries/who_standards/idn.pdf]. (assessed on 12 Dec. 2008)

21. ARIF G: Child Health and Poverty in Pakistan. The Pakistan Development Review 2004, 43(3):211–238.

22. Demographic and Health Survey: Bangladesh Demographic and Health Survey. Calverton, Maryland: BPS and ORC Macro; 2003.

23. Demographic and Health Survey: India Demographic and Health Survey. Calverton, Maryland: BPS and ORC Macro; 2005.

24. National surveillance project: Nutrition and Health surveillance in Barisal division. Nutritional Surveillance Annual report, Bulletin No. 7. Dhaka; 2002.

25. Martorell R, Habicht JP: Growth in early childhood in developing countries. Human growth: A comprehensive treatise. 2nd edition. Edited by: Falkner, JM. New York: Plenum Press; 1986:241–262.

26. Hoffman DJ, et al.: Energy expenditure of stunted and nonstunted boys and girls living in the shantytowns of Sao Paulo, Brazil. Am J Clin Nutr 2000, 72(4):1025–1031.

27. Prista A, et al.: Anthropometric indicators of nutritional status: implications for fitness, activity, and health in school-age children and adolescents from Maputo, Mozambique. Am J Clin Nutr 2003, 77(4):952–959.

28. Moestue H, Huttly S: Adult education and child nutrition: the role of family and community. J Epidemiol Community Health 2008, 62(2):153–159.

29. Harpham T, De Silva MJ, Tuan T: Maternal social capital and child health in Vietnam. J Epidemiol Community Health 2006, 60(10):865–871.

30. Addison CC: Nutrition and an active lifestyle. From knowledge to action. J Epidemiol Community Health 2006, 60(8):735.

Determinants of Use of Maternal Health Services in Nigeria — Looking Beyond Individual and Household Factors

Stella Babalola and Adesegun Fatusi

ABSTRACT

Background

Utilization of maternal health services is associated with improved maternal and neonatal health outcomes. Considering global and national interests in the Millennium Development Goal and Nigeria's high level of maternal mortality, understanding the factors affecting maternal health use is crucial. Studies on the use of maternal care services have largely overlooked community and other contextual factors. This study examined the determinants of maternal services utilization in Nigeria, with a focus on individual, household, community and state-level factors.

Methods

Data from the 2005 National HIV/AIDS and Reproductive Health Survey—an interviewer-administered nationally representative survey—were analyzed to identify individual, household and community factors that were significantly associated with utilization of maternal care services among 2148 women who had a baby during the five years preceding the survey. In view of the nested nature of the data, we used multilevel analytic methods and assessed state-level random effects.

Results

Approximately three-fifths (60.3%) of the mothers used antenatal services at least once during their most recent pregnancy, while 43.5% had skilled attendants at delivery and 41.2% received postnatal care. There are commonalities and differences in the predictors of the three indicators of maternal health service utilization. Education is the only individual-level variable that is consistently a significant predictor of service utilization, while socio-economic level is a consistent significant predictor at the household level. At the community level, urban residence and community media saturation are consistently strong predictors. In contrast, some factors are significant in predicting one or more of the indicators of use but not for all. These inconsistent predictors include some individual level variables (the woman's age at the birth of the last child, ethnicity, the notion of ideal family size, and approval of family planning), a community-level variable (prevalence of the small family norm in the community), and a state-level variable (ratio of PHC to the population).

Conclusion

Factors influencing maternal health services utilization operate at various levels—individual, household, community and state. Depending on the indicator of maternal health services, the relevant determinants vary. Effective interventions to promote maternal health service utilization should target the underlying individual, household, community and policy-level factors. The interventions should reflect the relative roles of the various underlying factors.

Background

Approximately 536,000 maternal deaths occur annually, of which over 95% occur in sub-Saharan Africa and Asia [1]. Africa has the highest burden of maternal mortality in the world and sub-Saharan Africa is largely responsible for the dismal maternal death figure for that region, contributing approximately 98% of the maternal deaths for the region [1]. The lifetime risk of maternal death in sub-Saharan Africa is 1 in 22 mothers compared to 1 in 210 in Northern Africa, 1 in 62 for Oceania, 1 in 120 for Asia, and 1 in 290 for Latin America and the Caribbean [1]. Nigeria is a leading contributor to the maternal death figure in sub-Saharan Africa not only because of the hugeness of her population but also because of her high maternal mortality ratio. Nigeria's maternal mortality ratio of 1,100 is higher than the regional average [2]. With an estimated 59,000 maternal deaths, Nigeria which has approximately two percent of the world's population contributes almost 10% of the world's maternal deaths [3].

Scientific evidence has clearly established the inverse relationship between skilled attendants at birth and the occurrence of maternal deaths. Thus, the considerable variation in the maternal mortality estimates between different locations within the same region can be attributed, to a large degree, to the differences in the availability of and access to modern maternal health services [3]. The use of maternal health services also contributes to neonatal health outcomes as the health of the mother and the newborn is closely linked. Maternal complications in labor, for example, carry a high risk of neonatal death [4,5]. Three-quarters of neonatal deaths occur in the first week, and the highest risk of death is on the first day of life. Furthermore, the main direct causes of neonatal death, globally, are preterm birth (28%), severe infections (26%), and asphyxia (23%) [5]. This epidemiological picture underscores the contribution of the delivery process to neonatal deaths.

While available evidence indicates limited benefit from traditional antenatal care services, focused antenatal care provides opportunity for early detection of diseases and timely treatment. It also provides opportunities for preventive health care services such as immunization against neonatal tetanus, prophylactic treatment of malaria through the use of intermittent presumptive treatment approach, and HIV counseling and testing. Furthermore, antenatal care exposes pregnant women to counseling and education about their own health and the care of their children. Thus, antenatal care may be particularly advantageous in resource-poor developing countries, where health seeking behavior is inadequate, access to health services is otherwise limited, and most mothers are poor, illiterate or rural dwellers. With the strong positive association that has been shown to exist between level of care obtained during pregnancy and the use of safe delivery care, antenatal care also stands to contribute indirectly to maternal mortality reduction [6]. According to the 2003 Nigeria Demographic Health Survey (NDHS) [7], 37% of women who had births within the five years prior to the survey received no antenatal care for their most recent delivery while only 35.2% were assisted at delivery by a skilled attendant.

Several studies have assessed the individual and household determinants of utilization of maternal services. These studies have not yielded a consistent pattern of relationships between service utilization and individual and household predictors. In some cases, even when a strong association has been reported, such as in the case of the positive relationship between education and the use of skilled health attendants at birth, the extent and nature of the relationship are not uniform across social settings. For example, whereas studies in Peru [8] and Guatemala [9] showed that women with primary level education were more likely to utilize maternal health services compared to those without any formal education, some studies in Thailand [10] and Bangladesh [11] did not record any significant difference between the two educational groups. Distances to health services and rural locations have been generally reported to be strongly and negatively associated with the use of maternal health services [6]. Some studies conducted in Turkey [12] and southern India [13,14], however, did not show any significant difference in the use of antenatal care between urban and rural women. Association between age and service utilization has also been inconsistent across studies. Whereas many studies found a positive correlation between age and the use of skilled attendants at child birth [12,15-18], others have found a curvilinear relationship [19,20]. Religion has also shown variable pattern of association with service utilization, with significant association in some settings [21] but not in some others [13]. In contrast, parity has been consistently shown to be negatively correlated with the use of skilled attendants [10,14,15,19,22]. A number of studies have reported positive association between economic status and use of medical settings for delivery [13,10,16] whereas others have not found such an association [23,24].

One important inference from the review of existing literature is that the role of individual and household factors differs from one geographic and social setting to another. Thus, as several authors have aptly noted, the determinants of maternal health care service utilization vary across and within cultures [13,25].

It is reasonable to assume that utilization of maternal health services depends on individual and household factors, as well as factors operating at the community or policy levels. The review of extant literature however shows that very few studies have gone beyond individual and household factors to consider factors at the community and higher levels. The implication of this omission is that some determinants are inadvertently missed, leaving a serious research and programmatic lacuna. Secondly, failure to consider the role of factors operating beyond the household level in service utilization may result in serious bias in the estimates. Individuals are nested within families, which are in turn nested within communities. Methodologically, it is important to take this nested structure into account.

This demands the use of multilevel modeling, which would calculate the standard errors more accurately and reduce the chance of misestimating the significance of variables, as some of the assumptions inherent in traditional regression methods are not valid for nested data [26].

Very few population-based studies have been carried out in Nigeria regarding determinants of maternal service utilization; most maternal health studies in the country have been institution-based. Most of the population-based studies were small-scale research, focusing on a handful of communities, usually small-sized rural communities [27-29]. Their geographic scope limits the applicability of their result on a large scale, particularly considering the complex multi-ethnic setting of Nigeria. In addition, most did not control for important confounding variables. Drawing from a nationally representative survey, this paper seeks to address the identified research gaps by examining the effect of individual, household, community and state-level factors on maternal care services utilization and employing strong analytical procedures. Specifically, we investigate the patterns and determinants of the utilization of the three dimensions of pregnancy-related care—ante-natal, delivery, and post-natal services.

Methods

Data

The data that we analyze in this paper derive from the 2005 National HIV/AIDS and Reproductive Health Survey (NARHS), a household survey designed to provide quantitative data for monitoring the impact of reproductive health interventions in Nigeria. The survey covered all the 36 states of Nigeria and participants were selected through a multi-stage probability sampling method. Details about the sampling have been provided elsewhere [30]. The original sample included 4,685 women (aged 15-49 years) and 5,396 men (aged 15-64 years); however, the sample included in the analyses reported in this paper was limited to the 2148 women who had a baby during the five years preceding the survey.

In addition to using data from the 2005 NARHS, we also accessed some state-level data published by National Bureau of Statistics for the year 2005 [31]. Specifically, we included information about the average number of residents to a Primary Health Care (PHC) facility in our estimated models. We use this variable as a proxy for the availability of maternal health services in the state.

Measurement

We analyze the predictors of three indicators of use of maternal health services: use of antenatal care, delivery assisted by a trained medical personnel (doctor or nurse/nurse-midwife), and use of postnatal care services. We assess the predictors of each of these indicators separately and with reference to the most recent birth.

We examined the predictive value of a number of individual and household variables, including rank of the most recent birth, education, ethnicity, age at last birth, attitudes towards family planning, ideal family size and socio-economic status. We examined the role of three community level variables: type of place of residence (urban versus rural), media saturation in the local government area (LGA) of residence, and prevalence of the small family norm in the LGA of residence. At the state level, we assessed the role of the ratio of Primary Health Care (PHC) facilities to the population. In addition, we assessed random effects at the state level. We selected these predictors based on information from extant literature and because they were significant predictors in initial bivariate analyses of the data. We describe the various predictors in Table 1.

Table 1. Measurement of various predictors included in the estimated models

Predictor	Measurement
Rank of most recent birth:	We distinguish between mothers whose most recent birth is rank 1 or 2 (34.7%) and those whose most recent birth is of a higher rank.
Education:	Highest level of education attained is divided into four categories: none, primary, secondary, and post-secondary education.
Ethnicity:	We specifically recognized the largest ethnic groups in Nigeria (Hausa, Igbo, Yoruba, Fulani and Kanuri) while all the other ethnic groups are classified together.
Age at last birth:	The questionnaire did not include a direct question on the age at last birth; we computed this indicator by subtracting the child's age from the woman's current age and rounding the result to the nearest whole number.
Attitudes towards family planning:	We measure this indicator through reported approval of family planning.
Ideal family size:	We distinguish between the women who gave a numeric response to the question on ideal family size (52.0%) and those who gave non-numeric responses, such as "Up to God" (48.0%).
Household socio-economic status:	We constructed a scale for household socio-economic status from information on possession of specific household items and utilities, including refrigerator, radio, television, car, video player, cell phone, standby generator, electricity, fan, kerosene stove, pipe-borne water and water closet (Cronbach's alpha for internal reliability: 0.88). The resulting scale was divided into five quintiles.
Urban residence:	This variable was derived from the question on the type of place of residence; we compare urban residents with their rural counterparts.
Media saturation in the LGA of residence:	We operationalize this community-level variable through the mean level of exposure to the radio and the television for the people in the LGA of residence other than the index individual (the non-self mean). We divide the measure into three categories, viz.: low, medium and high levels of community media saturation based on the percentiles.
Prevalence of small family norm in the LGA of residence:	We measured this variable using the non-self mean of expressed preference for a small family (four children or less).
State of residence:	The NARHS 2005 survey took place in the 36 states and the Federal Capital territory. The state of residence was included as a random variable in the estimated models to represent unmeasured factors related to the socio-political and cultural context.
Number of people per PHC in the state of residence:	This information came from the statistics published by the National Bureau of Statistics for the year 2005.

Analysis

Individuals are nested within households, households are nested within LGAs and LGAs are nested within states. In order to assess the roles of measured individual, household, community and state factors as well as unmeasured factors at the state level, we use multilevel modeling in this paper. The nature of nested data makes the use of traditional regression methods inappropriate: some of the assumptions inherent in traditional regression methods, including the assumption of independence among individuals within the same group and the assumption of equal variance across groups are not valid in the case of nested data [26].

We estimated a multilevel model that assessed the predictive values of measured individual, household, community and state factors (fixed effects) in addition to state-level random effects using the gllamm command in Stata [32]. For each of our three dependent variables, we estimated two models: an empty model that contains no covariates, and a full model that included fixed effects at the individual, household, community and state levels, and state-level random effects. The empty model allows us to verify if the magnitude of random effects at the state level justifies assessing random effects at that level. For all the estimated models, we evaluated the significance of the random effects using one-sided p-values rather than simple Wald tests since the null value is on the border of the parameter space [33,34].

Results

About three-fifths (60.3%) of the respondents used antenatal services at least once during their most recent pregnancy. The percentage of last births whose delivery was assisted by qualified medical personnel (doctor, nurse or nurse-midwife) was 43.4% while only two fifths (41.2%) received postnatal care (Table 2).

Bivariate Analysis

Table 2 shows variations in the three indicators of maternal health service utilization by selected socio-demographic, household, community and state factors. The results show that for each of the three indicators, there are significant differences by education, age at last birth, ethnicity, child's rank of birth, attitudes towards family planning, and ideal family size. There are also significant variations in the indicators by household socio-economic status, urban residence, community media saturation, prevalence of the small family norm in the LGA of residence, and the ratio of PHC to the population in the state of residence. For example, the three indicators of use increase steadily with education and household socio-economic status. In contrast, the indicators of use decrease by the child's rank of birth and are lower for women who gave non-numeric fertility ideals than their peers who reported gave numeric family size ideals. The relationship with age at last birth does not appear to be linear as the data show that the women most likely to use the antenatal care, medical personnel for delivery or postnatal care are those aged 25 - 34 years. In addition, the use of these services is more common among the women who approve of family planning compared to their peers who did not; urban women are also more likely to report use of the services compared to rural women. Yoruba, Igbo and minority women reported a significantly higher use of the services than Hausa, Fulani or Kanuri women.

Table 2. Variations in indicators of use of maternal and child health services, by selected individual, household and community characteristics.

Characteristics	n	Percent reporting use of:		
		Antenatal care	Medical personnel at delivery	Postnatal care
Individual Factors				
Education				
None	1013	37.1	17.7	19.3
Primary	537	74.3	51.5	49.3
Secondary	501	84.6	75.8	67.6
Post-secondary	107	95.3	93.4	83.2
Age at Last Birth				
15 – 19	351	46.7	29.0	28.8
20 – 24	546	58.2	37.5	36.1
25 – 29	546	69.2	53.5	50.3
30 – 34	367	68.1	50.9	49.0
35 – 39	218	56.4	43.6	40.8
40 +	130	52.3	43.1	36.1
Ethnic Group				
Hausa	656	37.5	15.2	18.0
Yoruba	313	91.4	81.4	70.9
Igbo	221	87.3	78.7	73.7
Fulani	136	41.9	17.6	27.2
Kanuri	66	27.3	22.7	24.2
Others	766	65.4	48.2	43.5
Child's rank of birth				
1 – 2	750	64.8	49.3	45.7
3 +	1408	57.9	40.2	38.8
Attitudes towards family planning				
Approve	1097	77.7	61.4	57.9
Disapprove	1061	42.2	24.7	23.9
Ideal family size				
Provided a numeric response	1122	74	61.0	55.1
Provided a non-numeric response (Up to God, etc.)	1036	45.4	24.3	26.0
Household Factors				
Household socio-economic status				
Very poor	440	27.9	13.2	17.0
Poor	466	44.0	24.9	22.3
Medium	394	62.4	39.8	38.8
Rich	431	80.0	64.7	58.4
Very rich	427	89.4	74.6	71.4
Community Factors				
Type of Place of Residence				
Rural	1495	49.3	32.1	31.4
Urban	663	84.9	68.8	63.3
Community media saturation				
Low	858	35.1	18.4	18.4
Medium	692	67.6	45.2	45.8
High	608	87.5	78.6	68.1
Prevalence of small family norm in community				
Low (0 – 10%)	927	35.7	15.6	19.6
Medium (11 – 30%)	663	68.5	48.7	45.1
High (< 30%)	568	90.8	82.6	71.8
State-level Factors				
Average number of people to a PHC in the state of residence				
Small (< 5500)	737	71.5	52.4	47.5
Medium (5500 – 9000)	705	60.0	46.8	44.2
Large (> 9000)	716	49.0	30.8	31.7
All Respondents	2158	60.3	48.4	44.8

Source: Nigeria National HIV/AIDS and Reproductive Health Survey, 2005; Data on the ration of people to a Primary Health Center (PHC) came from the 2006 Core Welfare Indicator Questionnaire Survey, a national survey conducted by the Nigeria Federal Bureau of Statistics.

Use of these services also increases steadily with community media saturation and the prevalence of the small family norm in the LGA of residence. In contrast, the three indicators decrease as the average number of people per PHC in the state of residence increases.

Multilevel Models

Obviously, the bivariate relationships indicated by the data on Table 2 can be due to interrelationships among the various measured characteristics as well as to unmeasured characteristics at the state level. We therefore used multilevel modeling to determine the predictors of maternal health services utilization and parse the variance in use into its fixed and random components. In the multilevel model, state of residence is modeled to be random.

We started with an empty, intercept-only model to test the null hypothesis that state-level variance in maternal health services utilization is zero and to assess if our data justify the decision to assess random effects at the state level. The results presented in Table 3 show that for each indicator of maternal health services utilization there is considerable between-states heterogeneity. For example, for antenatal care, the state-level variance in the empty model is large and significant pointing to considerable differences in use across states. The conditional intra-class (ICC) correlation in the empty model for antenatal care indicates that 36.8% of the total variance in use of antenatal care is attributable to the differences across states; in other words, use of service cluster significantly by state. We find similar results for assisted delivery and use of postnatal care (Table 3)

Table 3. Parameter coefficients for the multilevel model for various indicators of use of maternal and child health services—Empty model, no covariates.

	Antenatal care	Medical personnel at delivery	Postnatal care
Random Effects			
State Level Variance	1.92*** (0.50)	2.17*** (0.54)	1.67*** (0.44)
Rho – Intra-class correlation	0.368	0.397	0.337
Log-likelihood	-1165.64	-1125.421	-1221.41
AIC	2335.3	2254.8	2446.8

Source: Nigerian National HIV/AIDS and Reproductive Health Survey, 2005
Notes: *** p < 0.001;

We now turn our attention to the results of the full models that assess the roles of predictors at the various levels (Table 4). Since there are some differences in the predictors of the specific indicators of use, we present the findings separately for each indicator.

Table 4. Results of the multilevel analysis of the predictors of indicators of use of maternal and child health services.

Characteristics	n	Odds Ratio (Std. Error)[a]		
		Antenatal care	Medical personnel at delivery	Postnatal care
Fixed Effects[b]				
Individual Factors				
Education				
None (RC)	1013	1.00	1.00	1.00
Primary	537	1.88*** (0.30)	1.69*** (0.28)	1.65*** (0.25)
Secondary	501	2.01*** (0.43)	3.01*** (0.60)	2.06*** (0.38)
Post-secondary	107	5.03** (2.64)	10.68*** (4.88)	3.50*** (1.15)
Age at last birth in single years	2158	1.18** (0.07)	1.08 (0.07)	1.13* (0.07)
Square of age at last birth	2158	0.99** (0.001)	0.99 (0.001)	0.99* (0.001)
Ethnic Group				
Hausa (RC)	656	1.00	1.00	1.00
Yoruba	313	1.22 (0.51)	1.62 (0.58)	1.57 (0.52)
Igbo	221	2.09§(0.87)	3.76*** (1.49)	2.10* (0.78)
Fulani	136	0.68 (0.18)	0.77 (0.25)	1.22 (0.33)
Kanuri	66	0.67 (0.34)	1.72 (0.90)	0.97 (0.50)
Others	766	1.35 (0.32)	2.04*** (0.53)	1.55§(0.38)
Child's rank of birth				
3 + (RC)	750	1.00	1.00	1.00
1 – 2	1408	1.17 (0.19)	1.22 (0.20)	1.10 (0.16)
Attitudes towards family planning				
Disapprove (RC)	1097	1.00	1.00	1.00
Approve	1061	1.64*** (0.22)	1.28§(0.17)	1.50*** (0.20)
Ideal family size				
Provided a non-numeric response (Up to God, etc.) (RC)	1036	1.00	1.00	1.00
Provided a numeric response	1122	1.14 (0.16)	1.71*** (0.26)	1.47** (0.19)
Household Factors				
Household socio-economic status				
Very poor (RC)	440	1.00	1.00	1.00
Poor	466	1.53* (0.27)	1.88** (0.43)	1.01 (0.20)
Medium	394	2.40*** (0.48)	2.72*** (0.62)	1.69** (0.34)
Rich	431	3.76*** (0.87)	4.27*** (1.07)	2.46*** (0.55)
Very rich	427	5.06*** (1.69)	4.34*** (1.23)	3.02*** (0.76)
Community Factors				
Type of Place of Residence				
Rural (RC)	1495	1.00	1.00	1.00
Urban	663	2.36*** (0.48)	1.69** (0.33)	1.63** (0.29)
Community media saturation				
Low (RC)	858	1.00	1.00	1.00
Medium	692	1.51* (0.25)	1.44* (0.27)	1.51* (0.26)
High	608	1.29 (0.36)	2.17*** (0.58)	1.20 (0.30)
Prevalence of small family norm in community				
Low (< 11%) (RC)	927	1.00	1.00	1.00
Medium (11 – 30%)	663	1.39 (0.30)	1.50§(0.33)	1.14 (0.34)
High (< 30%)	568	1.91* (0.58)	1.85* (0.53)	1.49 (0.40)
State-level Factors				
Average number of people to a PHC in the state of residence				
Small (< 3500) (RC)	737	1.00	1.00	1.00
Medium (3500 – 9000)	735	0.63 (0.24)	0.71 (0.24)	0.87 (0.30)
Large (> 9000)	716	0.42* (0.16)	0.41** (0.14)	0.66 (0.23)
Random Effects				
State Level Variance		0.70** (0.23)	0.56** (0.19)	0.60** (0.20)
Residual intra-class correlation		0.183	0.152	0.160
Log-likelihood		-939.50	-881.86	-1052.80
AIC		1931.01	1815.73	2157.61

Source: Nigerian National HIV/AIDS and Reproductive Health Survey, 2005; Data on the ratio of people to a Primary Health Center (PHC) came from the 2006 Core Welfare Indicator Questionnaire Survey, a national survey conducted by the Nigeria Federal Bureau of Statistics.
Notes: § p < 0.1; * p < 0.05; ** p < 0.01; *** p < 0.001
RC = reference category
Standard errors are in parenthesis
[a]Model with fixed effects at the individual, household, community and state levels, and random effects at the state level;
[b]Fixed effects expressed as odds ratio

Antenatal Care

The most significant individual-level predictors of use of antenatal care services are education, age at the birth of last child, and attitudes towards family planning (Table 4). The odds of reporting use of antenatal care services increase steadily with education such that the women with post-secondary education are five times as likely to report service use as their counterparts with no formal education. Approval of family planning, a variable reflecting modern as opposed to conservative ways of thinking, is associated with a 64% increase in the odds of reporting use. The relationship between age at the birth of the last child and use of antenatal care services appears to be curvilinear. The negative coefficient associated with the square of age indicates that use of antenatal services initially increases with age up to a threshold and decreases thereafter.

Household socio-economic status is positively related with use of antenatal services such that the odds of reporting use are almost six times as high among women from the richest households compared to their counterparts from the poorest households. The three community level variables included in the model turn out to be significant predictors of antenatal services utilization. Living in an urban community increases the odds of antenatal service utilization more than twofold. The data show a rather curious relationship with community media saturation: compared to low media saturation, a high level of community media saturation does not seem to make a difference whereas medium level does. In contrast, concerning the small family norm, it appears that only a high level of prevalence makes a significant difference.

The data further show that the larger the number of residents served by a PHC in the state, the less the odds that women would use antenatal care services. Finally, state-level random effects are significant; the residual intraclass correlation is still appreciably large, indicating that even after controlling for individual, household and community factors, there is still considerable clustering of antenatal service utilization at the state level.

Skilled Assistance at Child Birth

The individual-level variables that predict use of skilled (medical) personnel for delivery include education, ethnicity, and family size ideals (Table 4). As we saw with the use of antenatal care, the use of medical personnel for delivery increases steadily with education. This indicator of service use is also a function of ethnicity: compared to women of Hausa descent, Igbo and minority women are significant more likely to report use of skilled assistance for delivery. Providing a numeric ideal family size is also associated with increased odds of using medical personnel at delivery.

Furthermore, the odds of reporting this indicator of use increase monotonically with household socio-economic status and are higher for urban residence compared to their rural peers. In addition, both community media saturation and the prevalence of the small family norm in the community present a graduated, dose-response relationship with use of medical personnel for delivery. State-level random effects are also significant and there is evidence of clustering at the state level even after controlling for individual, household and community variables.

Postnatal Care

At the individual level, education, age at the birth of the last child, ethnicity, approval of family planning and family size ideals are the strongest predictors of postnatal care (Table 4). Specifically, as we noticed with the two previous indicators, use of postnatal care increases consistently with education. The odds of using postnatal care are also significantly higher for women who approve of family planning and who report numeric family size ideals compared to their counterparts who did not report these attitudes. Although not as strong as what we observed for antenatal use, age presents a curvilinear relationship with use of postnatal care.

At the household level, socio-economic status is a significant positive predictor. Two community-level variables—urban residence and community media saturation—are significant predictors but the prevalence of the small family norm is not. Unlike what we observed for the two previous indicators of service utilization, the indicator of health services availability in the state does not appear to make a significant difference for use of postnatal care. Nonetheless, the state-level random effects are significant with unmeasured state-level factors accounting for 16% of total variance in the use of postnatal care.

Discussion

This study is based on the NARHS, which involved a nationally representative population sample, and marks a departure from most of the previously reported studies on maternal health services utilization in Nigeria in terms of its national coverage. In addition, unlike most previous studies, we covered the three dimensions of pregnancy-related care—antenatal, delivery and postnatal services.

Our results show that the level of utilization of orthodox health care facilities for maternal care among women in Nigeria is low. Indeed utilization of maternal health care services is lower in Nigeria than in many countries in sub-Saharan Africa. For example, whereas we found that 60.3% of Nigerian mothers utilized antenatal care services during their last birth, the comparative figures were

88.0% for Benin (2006 DHS), 72.8% for Burkina Faso (2007 DHS), 83.4% for Cameroon (2004 DHS), and 91.9% for Ghana (2003 DHS) [35].

Similarly, the indicators of skilled assistance during delivery and use of postnatal care are considerably lower in Nigeria than in most African countries. A recent UNICEF report [36] shows that regarding skilled assisted delivery, only Burundi, Chad, Eritrea, Ethiopia, Niger and Somalia performed more poorly than Nigeria in sub-Saharan Africa.

The finding that utilization of antenatal services is higher than use of skilled assistance during delivery is consistent with the results of previous studies conducted in Nigeria [29,37] and elsewhere [38-40]. One of the reasons that have often been advanced for the lower coverage of skilled and institutional delivery compared to antenatal care coverage is the unpredictable nature of the onset of labor in the face of difficulty in accessing health facilities in resource-poor environments. Many rural communities in sub-Saharan Africa are examples of such environments, with the characteristic poor road networks, limited transportation means and underserved population in terms of health facilities. Our study would support such an explanation considering that the average number of residents per PHC is a more significant predictor of use of skilled assistance for delivery than of use of antenatal care.

The poor staffing of the health facilities, particularly the primary health care facilities, which makes it difficult to guarantee 24-hour availability of services had also been reported as a factor that discourages women in Nigeria, even when they had received antenatal care services, to seek medical services when labor commences [41]. The role of traditional and religious beliefs as well as the perception of women with regards to comparative efficacy of the medical versus traditional birth attendants may also be contributory to failure to have skilled attendants at birth. As Addai [42] pointed out, modern (medical) and indigenous maternal health care services coexist in most African communities, particularly in rural areas, and women may have to choose between the two options. Some previous studies had reported that many Nigerian women, particularly those in rural areas, rate the services of the traditional birth attendants (TBAs) as being of higher quality than that of medical healthcare practitioners, particularly with regards to interpersonal communications and relationships [41,43]. TBAs have been reported to be more considerate and to provide more compassionate care. Women in rural Guatemala have similarly been reported as being less likely to deliver in medical settings because of lack of social support provided by health-care professionals compared with traditional midwives [23]. Furthermore, Falkingham [24] reported that despite the fact that medical services were accessible and free of charge, women in Tajikistan prefer to deliver at home because they perceive available medical services to be of low quality and unsafe. Economic reason also ranks strongly in the preference of some Nigerian women for TBAs as their services have been reported to be more affordable. Additionally, TBAs may offer a more convenient user-charges system that allows payment to be spread over a period of time or even to be made in kind [44].

Our finding regarding the significant positive association between education and each of the three indicators of maternal services use agrees with previous reports [8,12,45,46]. Education serves as a proxy for information, cognitive skills, and values; education exerts effect on health-seeking behavior through a number of pathways [47]. These pathways include higher level of health awareness and greater knowledge of available health services among educated women, improved ability of educated women to afford the cost of medical health care, and their enhanced level of autonomy that results in improved ability and freedom to make health-related decisions, including choice of maternal services to use [10,12,48]. Educated mothers are more likely to take advantage of public health-care services than other women [49,50]. Education may also impart feelings of self-worth and confidence as well as reduce the power differential between service providers and clients, thereby reducing the reluctance to seek care [51,52].

The absence of a statistically significant association between the child's rank of birth and maternal services utilization among Nigerian women is surprising. Previous studies have found a strong negative association between parity and maternal services utilization [38,46,53,54].

Our finding with regards to the association between ethnicity and service utilization is an interesting one. Whereas ethnicity seems to make no significant difference for use of antenatal care, it does for use of skilled assistance and postnatal care. For these two indicators of use, the Fulanis, and the Kanuris (in the north) are not statistically different from the Hausas (in the north). In contrast, the level of service utilization was significantly higher among the Igbos (in the south) and the "minority" tribes compared to the Hausas. The pattern is consistent with the general picture of wide regional disparity in health status in Nigeria's diverse and multi-ethnic setting as has been reflected, for example, in the NDHS [7]. Perhaps more than other factors, this result reflects the influence of culture. An analysis of the social context of childbirth among the Hausas of Northern Nigeria, for example, has highlighted the strong influence of cultural beliefs and practices on childbirth and related fertility-related behaviors, and their significant contribution to the maternal morbidity and mortality picture [55]. In addition to the fact that a high proportion of teenage girls are married out to much older men, sometimes as early as 9 or 10 years of age, based on religious/cultural beliefs, cultural norms restrict women from readily seeking health-related assistance in pregnancy and childbirth. As Wall [55] noted, "Kunya, or 'shame' plays an extremely important role in Hausa childbirth, particularly in the first pregnancy. The newly pregnant girl should not draw attention to her gravid state, and all mention of the pregnancy should be avoided in conversation and action. This social pressure to remain 'modest' may well prevent her from asking questions about childbirth, and creates a major barrier to her seeking skilled assistance for delivering in hospital. As Wall further note, the pregnant girl's "mother, other relatives, and a local midwife usually stay with her during labor, but her kunya and her fear may be so great that she does not say anything until labor is well advanced." (p. 353). If there is nobody immediately available, it is unlikely that the girl in labor will send for someone, as "kunya" will prevent her from saying anything. Moreover in the cultural context of the Hausas, delivering her first child alone—unattended to by anyone—is viewed with pride.

Whereas some previous Nigerian studies had reported a significant relationship between age and maternal services utilization [56,57], others had shown no such difference [28,58]. We found no significant relationship between age and use of skilled assistance. For the other two indicators, we found a curvilinear relationship, such that women in the middle childbearing ages are more likely to used maternal services compared to their peers in the early or late childbearing ages. This finding agrees with the report of Obermeyer and Potter [19] and Gage [20].

Expectedly, fertility-related attitudes, which are reflected in our study by attitudes towards family planning and notion about the ideal family size, have significant relationship with maternal health services utilization. Favorable attitudes towards family planning and a clear notion about what constitutes an ideal smaller family size reflect less conservative behavior and more openness to modern health-related concepts and services.

At the household level, we found socio-economic status to be a significant predictor: for each of the three indicators, use of maternal services increases steadily with socio-economic status. Studies elsewhere have also documented positive relationship between economic status and early antenatal care use [16,58,59], delivery in medical settings [13,14,16], and utilization of postnatal services [60,61].

A major focus of this study is to go beyond individual and household factors and investigate the effects of community and state level factors on maternal care services utilization. At the community level, urban residence was consistently associated with increased odds of service utilization. This finding is in consonance with previous studies which have reported a significantly higher use of services in urban compared to rural areas in Nigeria [7,22,43,62] and elsewhere [13,16,63,64]. The other two community factors assessed—community media saturation and the prevalence of the small family norm in the community—are expectedly significant predictors of service utilization. Note however that the small family norm was not significant for postnatal care. The reason for this finding is not clear.

At the state level, we found that the ratio of PHC to the population was a significant predictor for use of antenatal care and skilled attendance at delivery, but not for postnatal care. The relationships are such that the larger the number of residents to a PHC the less the odds of using the services. This negative relationship is understandable since the more people a PHC serves, the more likely it is that access to the services would be difficult and the quality of services received poor.

Finally, we found that the random effects of the state of residence on each of the three indicators of maternal care service utilization are significant. Substantively, this finding shows that unmeasured factors operating at the state of residence level play a significant role in determining utilization of maternal health services beyond the influence of individual, household and community factors.

The findings from this study have implications for evidence-based programming. Collectively, the findings highlight the need for programs to adopt a multilevel approach and address the factors affecting maternal health services utilization at various levels—individual, household and community. More specifically, programs need to explore effective ways of increasing service utilization among lowly educated and poor women who are the least likely to use maternal health services. Evidence from elsewhere have shown that access to services and cost are serious barriers to service utilization among the poor [65,66]. As Fotso et al. surmised, it is not enough to increase the availability of services, making such services affordable to the poor is a necessity [67].

The strong role of community-level normative factors point to the need for interventions that target social norms. For example, using the media to disseminate consistent messages promoting the use of maternal health services could help to increase discussion of these issues within the community, a relevant step towards changing prevailing negative norms. Also relevant are efforts that involve community leaders and other key persons as agents of change. The findings that the prevalence of the small family norm in the community and personal fertility-related attitudes are associated with differences in service utilization suggest that promoting the use of family planning may ultimately help to foster the utilization of other maternal health services. In other words, programs that seek to promote the small family norm and change attitudes that are unfavorable towards family planning are relevant.

The significant state-level random effects that our study found demonstrate the need to contextualize efforts aimed at promoting maternal service utilization. There are obviously some unmeasured factors at the state level that predict service utilization. An effective strategy should be state-specific and seek to identify and address state-level factors that affect service utilization.

This study has some limitations that should be noted. First, the NARHS, the source of data for our study, was based on the self-report of respondents, and provided no validation of obtained information with any objective source such as health facility cards. The validity of self-reported behavior constitutes a concern in the literature, but it is logical to assume that biases are less likely in pregnancy-related events as compared to sensitive issues such as sexual behavior and drug abuse. Social desirability bias may also be an issue in cases that women feel they need to respond in a way expected of them. The comparability of our results with that of NDHS with regards to antenatal care use, for example, suggests that such bias is not likely to have affected our findings in any significant way.

Second, some known predictors of service utilization are obviously missing from our analyses. For example, availability of maternal health services within the immediate locality of respondents and the distance of respondents to such health services could have contributed to the picture of utilization pattern. Unfortunately these variables were not available in the NARHS. Although we included the state-level measure of PHC density (the number of residents to a PHC) in our analyses, the extent to which this variable is a good proxy for individual-level variable is uncertain.

Third, the study relied on cross-sectional data with the attendant potential selectivity and endogeneity bias. There is a possibility that the relationships that we found in our study are due to the influence of unmeasured individual and community-level variables that are associated with both the dependent and independent variables in our estimated models. It is also possible that the observed relationships reflect reverse causation or are due to measurement error. There are analytic methods to adjust for endogeneity bias in cross-sectional data (e.g., propensity score matching, bivariate probit regression, multivariate probit regression and instrumental variable regression) [68]. Nonetheless, adjusting for endogeneity is beyond the scope of this paper.

Conclusion

Factors influencing maternal health services utilization operate at various levels—individual, household, community and state. While education, socio-economic level, and urban residence are consistently strong predictors of all the maternal health services considered in this study, other determinants of service utilization generally vary in magnitude and level of significance by the type of maternal service—ante-natal care, skilled attendant at birth, and postnatal care. To be optimally effective, interventions to promote maternal health service utilization need to take these findings into consideration: they should target the underlying individual, household, community and state-level factors that are relevant to each type of maternal health service. It is particularly important for interventions to explore effective ways of increasing service utilization among lowly educated and poor women in rural areas who are the least likely to use maternal health services.

Competing Interests

The authors declare that they have no competing interests.

Authors' Contributions

SB and AF were equally responsible for designing the study and drafting the manuscript. SB performed the statistical analyses while AF contributed the literature review.

Acknowledgements

The authors acknowledge the Society for Family Health (Nigeria) for granting us permission to use the 2005 NARHS data for this study. Financial assistance for the survey came from the United Kingdom Department for International Development (DFID) and the United States Agency for International Development (USAID). The Society for Family Health and the Nigeria Population Commission (NPC) provided technical assistance in the design and implementation of the survey.

References

1. World Health Organisation (WHO): Maternal mortality in 2005: estimates developed by WHO, UNICEF, UNFPA, and the World Bank. Geneva, WHO; 2007.

2. Hill K, Thomas K, AbouZahr C, Walker N, Say L, Inoue M, Suzuki E: Estimates of maternal mortality worldwide between 1990 and 2005: an assessment of available data. Lancet 2007, 370, 9595:1311–1319.

3. Federal Ministry of Health (FMOH) [Nigeria]: Road map for accelerating the attainment of the millennium development goals related to maternal and newborn health in Nigeria. Abuja, FMOH; 2005.

4. Kusiako T, Ronsmans C, van Der PL: Perinatal mortality attributable to complications of childbirth in Matlab, Bangladesh. Bull World Health Organ 2000, 78:621–27.

5. Lawn JE, Cousens S, Zupan J: Lancet Neonatal Survival Steering Team. 4 million neonatal deaths: when? Where? Why? Lancet 2005, 365(9462):891–900.

6. Bloom SS, Wypij D, Gupta M: Dimensions of women's autonomy and the influence on maternal health care utilization in a north Indian city. Demography 2001, 38:67–78.

7. National Population Commission (NPC) [Nigeria] and ORC Macro: Nigeria Demographic and Health Survey 2003. Calverton, Maryland: NPC and ORC Macro; 2004.

8. Elo IT: Utilisation of maternal health-care services in Peru: the role of women's education. Health Transit Rev 1992, 2:49–69.

9. Goldman N, Pebley AR: Childhood immunization and pregnancy related services in Guatemala. Health Transit Rev 1994, 4:29–44.

10. Raghupathy S: Education and the use of maternal health care in Thailand. Soc Sci Med 1996, 43:459–71.

11. Dharmalingam A, Hussain TM, Smith JF: Women's education, autonomy and utilization of reproductive health services in Bangladesh. In Reproductive health: programmes and policy changes post-Cairo. Edited by: Mundigo AI. Liege, Belgium, International Union for the Scientific Study of Population (IUSSP); 1999.

12. Celik Y, Hotchkiss DR: The socioeconomic determinants of maternal health care utilization in Turkey. Soc Sci Med 2000, 50:1797–1806.

13. Navaneetham K, Dharmalingam A: Utilisation of maternal health care services in southern India. Soc Sci Med 2002, 55:1849–1869.

14. Bhatia JC, Cleland J: Determinants of maternal care in a region of south India. Health Transit Rev 1995, 5:127–142.

15. Obermeyer CM: Culture, maternal health care and women's status: a. comparison of Morocco and Tunisia. Stud Fam Plann 1993, 24:354–365.

16. Gertler P, Rahman O, Feifer C, Ashley D: Determinants of pregnancy outcomes and targeting of maternal health services in Jamaica. Soc Sci Med 1993, 37:199–211.

17. Pebley AR, Goldman N, Rodríguez G: Prenatal and delivery care and childhood immunization in Guatemala: do family and community matter? Demography 1996, 33:231–247.

18. Magadi MA, Agwanda AO, Obare FO: A comparative analysis of the use of maternal health services between teenagers and older mothers in sub-Saharan Africa: evidence from Demographic and Health Surveys (DHS). Soc Sci Med 2007, 64:1311–1325.

19. Obermeyer CM, Potter JE: Maternal health care utilization in Jordan: a study of patterns and determinants. Stud Fam Plann 1991, 22:177–187.

20. Gage A: Premarital childbearing, unwanted fertility and maternity care in Kenya and Namibia. Population Studies 1998, 52:21–34.

21. Gyimah SO, Takyi BK, Addai I: Challenges to the reproductive-health needs of African women: on religion and maternal health utilization in Ghana. Sos Sci Med 2006, 62:2930–2944.

22. Ekele BA, Tunau KA: Place of delivery among women who had antenatal care in a teaching hospital. Acta Obstet Gynecol Scand 2007, 86:627–630.

23. Glei DA, Goldman N: Understanding ethnic variation in pregnancy-related care in rural Guatemala. Ethn Health 2000, 5:5–22.

24. Falkingham J: Inequality and changes in women's use of maternal health care services in Tajikistan. Stud Fam Plann 2003, 34:32–43.

25. Bashour H, Abdulsalam A, Al-Faisal W, Cheikha S: The patterns and determinants of maternity care in Damascus, Syria. East Mediterr Health J 2008, 14:595–604.

26. Guo G, Zhao H: Multilevel modeling for binary data. Annual Review of Sociology 2000, 26:441–462.

27. Okafor CB: Availability and use of services for maternal and child health care in rural Nigeria. Int J Gynaecol Obstet 1991, 34:331–46.

28. Nwakoby BN: Use of obstetric services in rural Nigeria. J R Soc Health 1994, 114:132–1366.

29. Osubor KM, Fatusi AO, Chiwuzie JC: Maternal health-seeking behavior and associated factors in a rural Nigerian community. Matern Child Health J 2006, 10:159–69.

30. Federal Ministry of Health (FMOH) [Nigeria] (2005): Technical Report, National HIV/AIDS and Reproductive Health Survey. 2005.

31. National Bureau of Statistics (n.d): [http://www.nigerianstat.gov.ng/social_statistics/SSD%20final.pdf]. Social Statistics in Nigeria. Abuja, Nigeria: National Bureau of Statistics; 2005.

32. Rabe-Hesketh S, Skrondal A, Pickles A: Reliable estimation of generalized linear mixed models using adaptive quadrature. Stata Journal 2002, 2:1–21.

33. Raudenbush SW, Byrk AS: Hierarchical linear models: applications and data analysis methods. 2nd edition. Sage, Thousand Oaks, CA; 2002.

34. Rabe-Hesketh S, Skondral A: Multilevel and longitudinal modeling using Stata. 2nd edition. Stata Press, College Station, TX; 2008.

35. Macro International Inc: [http://www.measuredhs.com/] MEASURE DHS STATcompiler. Accessed December 19, 2008

36. United Nations Children's Fund (UNICEF): The State of the World's Children 2008. Statistical Tables: 8: Women. New York, UNICEF; 2007:142–145.

37. Adekunle C, Filippi V, Graham W, Onyemunwa P, Udjo E: Patterns of maternity care among women in Ondo States, Nigeria. In Determinants of health and mortality in Africa. Edited by: Allan G Hill. Demographic and Health Survey Further Analysis Series. New York: The Population Council; 1990:1–45.

38. Mekonnen Y, Mekonnen A: Factors influencing the use of maternal healthcare services in Ethiopia. J Health Popul Nutr 2003, 21:374–382.

39. Stewart MK, Stanton CK, Ahmed O: Maternal health care. In DHS Comparative Studies. Calverton, Maryland, Macro International Inc; 1997.

40. Leslie J, Gupta GR: Utilization of formal services for maternal nutrition and health care. Washington DC, International Center for Research on Women; 1989.

41. Fatusi AO, Ijadunola KT: National Study on Essential Obstetric Care Facilities in Nigeria. Abuja, Federal Ministry of Health & United Nations Population Fund (UNFPA); 2003.

42. Addai I: Determinants of use of maternal-child health services in rural Ghana. J Biosoc Sci 2000, 32:1–15.

43. Fatusi AO, Abioye-Kuteyi EA: Traditional birth attendants in Nigeria: what do we know about them? In Defining the role of Traditional Birth Attendants in Nigeria. Proceedings of a National Workshop on the Roles of TBAs in Reproductive Health. Lagos, United Nations Population Fund (UNFPA); 1998:18–25.

44. Onah HE, Ikeako LC, Iloabachie GC: Factors associated with the use of maternity services in Enugu, southeastern Nigeria. Soc Sci Med 2006, 63:1870–1878.

45. Becker S, Peters DH, Gray RH, Gultiano C, Black RE: The determinants of use of maternal and child health services in Metro Cebu, the Philippines. Health Transit Rev 1993, 3:77–89.

46. Stewart K, Sommerfelt AE: Utilization of maternity care services: A comparative study using DHS data. Proceedings of the Demographic and Health Surveys World Conference. Washington, DC 1991, III:1645–1668.

47. Schultz TP: Studying the impact of household economic and community variables on child mortality. Population and Development Review 1984, 10:215–235.

48. Caldwell JC: Maternal education as a factor in child mortality. World Health Forum 1981, 2:75–78.

49. Orubuloye IO, Caldwell JC: The impact of public health services on mortality: a study of mortality differentials in a rural area of Nigeria. Population Studies 1975, 29:259–272.

50. Caldwell J: Education as a factor in mortality decline: an examination of Nigerian data. Population Studies 1979, 33:395–413.

51. Chanana K: Education attainment, status production and women's autonomy: a study of two generations of Punjabi women in New Delhi. In Girls' Schooling, Women's Autonomy and Fertility Change in South Asia. Edited by: Jeffery R, Basu AM. New Delhi, Sage Publications; 1996:107–132.

52. Starrs A, ed: The Safe Motherhood Action Agenda: Priorities for the Next Decade. New York: Family Care International; 1998.

53. Akin A, Munevver B: Contraception, abortion and maternal health services in Turkey: Results of further analysis of the 1993 Turkish Demographic and Health Survey. Calverton, Maryland, Ministry of Health [Turkey] and Macro Inc.; 1996.

54. van Eijk AM, Bles HM, Odhiambo F, Ayisi JG, Blokland IE, Rosen DH, Adazu K, Slutsker L, Lindblade KA: Use of antenatal services and delivery care among women in rural western Kenya: a community based survey. Reproductive Health 2006, 3:2.

55. Wall LL: Dead Mothers and injured wives: the social context of maternal morbidity and mortality among the Hausa of Northern Nigeria. Stud Fam Plann 1998, 29:341–359.

56. Ikeako LC, Onah HE, Iloabachie GC: Influence of formal maternal education on the use of maternity services in Enugu, Nigeria. J Obstet Gynaecol 2006, 26:30–34.

57. Adeoye S, Ogbonnaya LU, Umeorah OU, Asiegbu O: Concurrent use of multiple antenatal care providers by women utilising free antenatal care at Ebonyi State University Teaching Hospital, Abakaliki. Afr J Reprod Health 2005, 9:101–106.

58. Burgard S: Race and pregnancy-related care in Brazil and South Africa. Soc Sci Med 2004, 59:1127–1146.

59. McCaw-Binns A, La Grenade J, Ashley D: Under-users of antenatal care: a comparison of non-attenders and late attenders for antenatal care with early attenders. Soc Sci Med 2007, 40:1003–1012.

60. Chakraborty N, Islam MA, Chowdhury RI, Ban W: Utilisation of postnatal care in Bangladesh: evidence from a longitudinal study. Health Soc Care Community 2002, 10:492–502.

61. Dhakal S, Chapman GN, Simkhada PP, van Teijlingen ER, Stephens J, Raja AE: Utilisation of postnatal care among rural women in Nepal. BMC Pregnancy and Childbirth 2007, 7:19.

62. Federal Ministry of Health (FMOH) [Nigeria]: 2003 National HIV/AIDS & Reproductive Health Survey. Abuja, FMOH; 2003.

63. Magadi M, Diamond I, Nascimento Rodrigues N: The determinants of delivery care in Kenya. Soc Biol 2000, 47:164–88.

64. Stupp W, Macke BA, Monteith R, Paredez S: Ethnicity and the use of health services in Belize. J Biosoc Sci 1994, 26:165–77.

65. Onah HE, Ikeako LC, Iloabachie GC: Factors associated with the use of maternity services in Enugu, southeastern Nigeria. Soc Sci Med 2006, 63:1870–1878.

66. Amooti-Kaguna B, Nuwaha F: Factors influencing choice of delivery sites in Rakai district of Uganda. Soc Sci Med 2000, 50:203–213.

67. Fotso JC, Ezeh A, Oronje R: Provision and use of maternal health services among urban poor women in Kenya: What do we know and what can we do? J Urban Health 2008, 85:428–442.

68. Babalola S, Kincaid DL: New methods for estimating the impact of health communication programs. Commun Methods Meas 2009, 3:61–83.

"I Washed and Fed My Mother before Going to School": Understanding the Psychosocial Well-Being of Children Providing Chronic Care for Adults Affected by HIV/AIDS in Western Kenya

Morten Skovdal and Vincent O. Ogutu

ABSTRACT

With improved accessibility to life-prolonging antiretroviral therapy, the treatment and care requirements of people living with HIV and AIDS resembles that of more established chronic diseases. As an increasing number of people living with HIV and AIDS in Kenya have access to ART, the primary caregivers of poor resource settings, often children, face the challenge of meeting the requirements of rigid ART adherence schedules and frequent relapses. This, and the long-term duty of care, has an impact on the primary caregiver's experience of this highly stigmatised illness—an impact that is often described in relation to psychological deprivation. Reflecting the meanings attached to caregiving by 48 children in Western Kenya, articulated in writing, through photography and drawing, individual and group interviews, this paper presents three case studies of young caregiving. Although all the children involved in the study coped with their circumstances, some better than others, we found that the meanings they attach to their circumstances impact on how well they cope. Our findings suggest that only a minority of young caregivers attach either positive or negative meanings to their circumstances, whilst the majority attaches a mix of positive and negative meanings depending on the context they are referring to. Through a continuum of psychosocial coping, we conclude that to provide appropriate care for young carers, health professionals must align their understanding and responses to the psychosocial cost of chronic care, to a more nuanced and contextual understanding of children's social agency and the social and symbolic resources evident in many African communities.

Introduction

"My caring experiences make me feel happy, they will help me in the future. If I am left alone, I will be able to do the duties. I become strong; I don't become a weak child." Joyce, age 12

In a globalised world, the management and support of people living with AIDS is an issue that concerns us all [1]. With our growing understanding of HIV and AIDS and improved access to antiretroviral therapy (ART), the AIDS epidemic is amenable to intervention and treatment management, changing its course to mirror the disease processes of other chronic illnesses [2]. According to Thorne [3] chronic illnesses are long term, and require careful management and adjustment by the patients and their caregivers as the person with the disease may fluctuate between chronic and acute episodes. People living with HIV/AIDS also require careful management to sustain their health. This is particularly the case of those on ART whose diet, adherence to rigid treatment plans and psychosocial wellbeing has to be managed, a responsibility that is often shared with their primary caregiver.

Although ART coverage in sub-Saharan Africa increased by 33% in 2007, with 2.1 million people receiving ART [4], resulting in a decline of HIV/AIDS related mortality [5], numerous external factors impact on the success of ART, particularly in rural Africa [6]. It is therefore important to understand the action and decision processes that impact on ART adherence [7] and home-based care of people with HIV/AIDS [8]. Poor infrastructure and opportunity costs have meant that many people still go without life-improving medicines [9] and many of those who receive ART are unable to adequately adhere to the strict treatment plans [10,11]. A study in Tanzania reveals that fear of stigma and discrimination, additional costs of transportation, and supplementary food and negative associations with hospital staff, have deterred HIV infected people from following up on referrals for ART [12]. Non-adherence can result in relapses and drug resistance, and recent findings from Malawi suggest that poor compliance to ART can even result in increased mortality [13].

These difficulties highlight the importance of understanding the implications for long-term care of people living with AIDS. Primary caregivers are at the forefront, struggling with rigid ART adherence schedules and frequent relapses. Nevertheless, primary caregivers continue to play a crucial role in facilitating the adherence of treatment plans and the provision of economic, nursing and moral support [8,14]. Perhaps unsurprisingly, family caregiving is often perceived to be a burden on the caregivers and they are increasingly reported to be at risk of poor health and deprived psychosocial well-being [15,16]. The majority of research on informal caregivers of people living with AIDS is focused on women [8,17], however, growing attention is being given to young caregivers. In this article we use the term 'young carers' to refer to children under the age of 18 who provide nursing care and support for sick, disabled or elderly relatives or guardians affected by AIDS on a regular basis and play a key role in sustaining the household.

Research on young carers in Africa is still in its early stages and has so far been limited to the context of AIDS. Elsbeth Robson and colleagues have been in the forefront, identifying the circumstances that characterise young caregiving in Africa. They have explored their caring arrangements [18], their duties and responsibilities and how this negatively impact on their school attendance [19], the socioeconomic and structural influences that induce young caregiving in Africa [20,21] as well as the ethical implications of doing research with young carers [22].

Robson and colleagues have been cautious not to export western conceptualisations of young caregiving to the African context [18], and have highlighted the reported benefits of young caregiving [19]. Evans and Becker [23] in their recent comparative study of children caring for parents with HIV and AIDS in Tanzania and UK, approached children as social actors and usefully identified some of the social determinants that facilitate the resilience of caregiving children. Nevertheless, the needs and vulnerabilities that do characterise young carers, together with a predominant focus on the ill-effects of caregiving in the international literature on young carers [24,25], and emerging trends on exploring the psychological distress of children affected by AIDS [26-33], has encouraged a focus on the psychological well-being of young carers, usually starting with the assumption that caring is inherently a source of mental ill-health for young people.

As with their adult counterparts, children who provide care and have domestic responsibilities have been associated with fragile mental health [16,22,34-39]. In a comparative study between Zimbabwe and the United States, Bauman and colleagues [34,35] interviewed a total of 100 ill mothers living with HIV and AIDS and one of their children. By choosing to use depression scales, the study begins with the assumption that caring for an ill parent is an inherently traumatic experience that automatically puts children at risk of mental health problems. While this undoubtedly is true in some cases, we believe that this assumption reflects a Western mental health discourse and dominant representations of childhood as a period of innocence and mental fragility in the absence of adult protection.

Although a debate on the psychosocial well-being of young carers is imperative, we believe a different approach is required to provide meaningful psychosocial support to young carers. This approach can lead to a more profound understanding and knowledge of coping and well-being and provide us with a good starting point for moving toward better health. Such an approach has been usefully theorised by Antonovsky [40] who, through his theories of salutogenesis (latin for the origins of health) and 'sense of coherence,' argues that the meanings given by people in difficult circumstances to their life situations shapes their sense of coherence, which then impacts on how they cope with their circumstances [40,41]. In taking a salutogenic approach we are not seeking to develop specialised techniques or suggestions about how outside professionals can cure the stress and hardship faced by young carers through expert techniques such as psychotherapy. Rather, we seek to report on existing indigenous life strategies—developed by young carers themselves within their immediate communities—that facilitate the sense of coherence they construct for their lives and promote their movement toward coping and well-being. Meanings ascribed to stressful life events have previously shown to be critical in coping and promoting psychosocial well-being [42,43].

In line with Antonovsky's salutogenic approach and to broaden our understandings of the psychosocial well-being of young caregiving, this paper presents some of the characteristics that exemplify the circumstances of young caregivers, including the social resources available to them, and the meanings that they attach to caring. In doing so, the paper aims to outline their negotiation with local understandings of childhood as a time of duty and service through their cognitive ability to attach a meaning to their caring experiences and construct a positive identity around these meanings.

We began this paper by quoting a 12-year-old girl who spoke of her caring role as a source of strength, even happiness. This counterintuitive claim calls for an understanding of the psychosocial nature of the demands of child caregiving. In this paper we therefore seek to explore the psychosocial well-being of children providing care for people chronically ill from AIDS, suggesting that the psychosocial well-being of young carers is best promoted with a nuanced understanding of the circumstances that surround young caregiving.

Methodology

This paper reports on the first phase of a two-year action research project with young caregivers in Western Kenya. This qualitative study was granted clearance from the LSE ethics committee and the Department for Gender and Social Services in Kenya.

Setting

The project is located in Bondo district along the shores of Lake Victoria. Poverty is a major challenge for people in Bondo district. With 68.1% of the 261,000 people living in the district living in absolute poverty, Bondo is one of the poorest districts in Kenya [44]. Bondo also has one of the highest HIV prevalence rates in Kenya. Estimates from the 2002 district development plan [45] put the HIV prevalence rate in Bondo at 30% while a more recent and conservative figure estimates it to be 13.7%, still double the national average [46]. Until 1999, the Bondo area was part of the Bondo-Siaya district. Throughout the early stages of the HIV and AIDS epidemic, it was marginalised due to the poor reach of health services that concentrated in the Siaya area, presumably contributing to its exceptionally high HIV prevalence rates. Another explanation for high HIV prevalence rates is found in its geographical location. Bordering Lake Victoria, the fishing villages provide employment for fishermen and truck drivers from all over Kenya whose migration and constant movement have contributed to the spread of HIV and AIDS. Numerous international NGOs have since been established in the Bondo area in an effort to halt the spread of AIDS and seeking to promote home based care and orphan care and support. With ARVs freely available in Kenya, an increasing number of people have commenced antiretroviral therapy. Currently, an estimated 40% of those infected with HIV and AIDS in Kenya receive ART [4,47]. Whilst this positive development also holds true in Bondo, the opportunity costs and poverty characterising the district undermine the effectiveness of

ART. A combination of the distance and costs of travelling to the nearest ARV health facility, stigma, and an inadequate diet all contribute to frequent relapses and the continued need for care. Relapses, and the fact that 60% of people living with HIV and AIDS are still without ART [4] mean that many children in Bondo district have had their lives affected by AIDS, often providing chronic care and support to those affected by the disease. The two participating rural communities in Bondo district are situated in areas characterised by high HIV prevalence rates and research from a neighbouring division suggests that one out of three children below 18 years of age have lost at least one biological parent, and one out of nine have lost both biological parents [48]. Although Kenya has a relatively low national HIV and AIDS prevalence rate (6.7%) compared to other Southern African countries, we believe Bondo district is representative of many rural areas in some of the hardest hit countries in sub-Saharan Africa.

Data Collection and Analysis

In this paper we present three case studies in order to map out the types of experiences reported by children, and to locate them on what we will call a 'continuum of coping.' These case studies represent the end point of a lengthy and stage-wise process of data collection and analysis that we will outline in this section.

Our data collection involved photography, individual and group interviews involving 48 young caregivers aged 11 to 17. Approaching the children as experts on their own lives [49], multiple methods were used to gather the data to ensure all children had an opportunity to communicate about their experiences in a way that felt comfortable [50]. Adapting the photovoice methodology and process developed by Wang and colleagues [51-53], the generation of photovoice data involved four stages. The first stage was that of photo-taking. Over a two-week period children took photos guided by the following four questions: 1) 'What is your life like?,' 2) 'What is good about your life?,' 3) 'What makes you strong?' and 4) 'What needs to change?' The second stage involved getting children to choose six of their favourite photographs, encouraged to identify a mix of photos showing how they get by, things they lack and/or something that is important to them. In the third stage, the children reflected on their chosen photographs and wrote down their thoughts prompted by the following questions: 1) 'I want to share this photo because...,' 2) 'What's the real story this photo tells?,' and 3) 'How does this story relate to your life and/or the lives of people in your neighbourhood?' If the children wanted to share a story that they were unable to capture on camera (e.g. for ethical or practical reasons), they were encouraged to draw the situation. This exercise generated a total of 184 photos and 56 drawings, each accompanied by a written reflection/story. To further explore the findings generated from their written reflections, 24 individual interviews and two group discussions were conducted with the children.

This paper is part of a much wider study, which yielded, amongst other things, a 6 theme analysis of key themes structuring children's accounts of their experience: 1) dynamics and characteristics of luo society; 2) characteristics and perceptions of caring children in Western Kenya; 3) determinants of caring experiences; 4) social resources; 5) action-based coping and 6) psychology-based coping [cf. [54]]. These themes cut across individual accounts and reflect general representational resources identified across the 48 research participants. In this particular paper we focus on three individual life stories, presenting the stories of three children in ways that highlight the different ways in which children used these representational resources to give meaning to their lives, and how different life experiences/access to resources and supported to varyingly positive, negative or mixed evaluations of their caring experiences (see Figure 1).

Figure 1. Diagram of Meanings attached to caring.

We have also taken the liberty of abridging some of the more detailed narratives into poems, which summarise the content of their accounts of their lives in a way that transcends the narrative chronology whilst bringing forward their meanings, using the children's own words [55,56].

Findings

All children managed their caring responsibilities and coped despite adverse circumstances. However, different meanings attached to caring suggest that the psychosocial cost of caregiving should be viewed on a continuum of coping. As diagram 1 indicates, most of the children surrender to the circumstances prescribed to them and 'get on' with the challenge of providing care. They may attach both positive and negative meanings to their experiences depending on the context and circumstances. However, a minority of children saw caring either as a relief, something valuable they would not be without, or as something which has caused damage to their life. Through the case studies of Samuel, Carolyne and Pascal (pseudonyms have been used to protect the identity of the children) we present three different scenarios of life as a young caregiver and how these circumstances may contribute to the meanings they attach to their circumstances.

The meanings that young carers attach to their circumstances are influenced by the social environment. In particular local understandings of childhood were found to impact their sense of coherence. Carolyne, during an interview, explained that "the duty of a child is to help parents. A child is called a helper." Concurrent to the local perception of childhood as a time of duty and service is a sometimes conflicting understanding of childhood, which is more rights-based and stresses the importance of education. It is common for children and adults in Bondo to draw on both representations of childhood, depending on the context, and they should therefore not be seen as binary, but located on a dynamic continuum. Nevertheless, as the case studies will illustrate, it is against these understandings of childhood that young carers give meaning to their circumstances as they manage their caregiving, head-of-household duties and education.

Samuel, Age 13 (Positive Meaning)

Samuel was 9 years old when he first realised that his father was ill. He noticed his father's swollen hands and joints and explained this in terms of a spell having been cast on his father by someone who was jealous of his job. Soon after, "another spell was cast by another person on his legs" and he was bedridden. Although he was taken for prayers, his condition never changed. He had sores all over the body and Samuel applied creams to his body, washed him, gave him drugs, prepared food and spoon-fed him. Although his mother was around to help, she too got sick and was able to offer less and less support. Samuel's father died in 2005 when Samuel was 11 years old and he currently cares for his mother. Samuel is not caring for his mother alone. His little sister also helps by fetching firewood and water, and assists Samuel where ever she can. Samuel also has a supportive extended family network. His aunt has moved in to help them out. Aware of this support, Samuel's assessment of his situation allows him to reflect further on local understandings of childhood as a time of duty and service and his perceived identity and role as a young carer. These reflections led him to move in with his ageing neighbour who had been deserted by his family: "I went and lived with him, helping him out, cooked for him, fetched water, took care of the poultry and ensured he was clean." Rather than concentrating on the care of his mother, Samuel's decision to move in with his ageing neighbour not only reflects his commitment to be a young carer but also his assessment of his circumstances as low risk.

Samuel has always played an important role in sustaining the households in which he has lived, drawing on the resources available to him through his parents. He cultivates land with sorghum, some of which he sells to buy basic amenities or chicken or goats. "Chicken help me in providing eggs, meat and gives us money if we sell them [...] at times I sell a chicken to get school fees." Samuel's family also has a cow whose milk can pay for his school uniform and which can be slaughtered for a funeral. The resources available to Samuel, combined with his active participation, allows Samuel to cope. From a psychosocial perspective, the availability of social resources and support makes it easier for Samuel to actively identify himself with childhood as a period of duty and service, successfully exceeding local expectations of childhood and from that create a positive caring identity. This positive identity came out strongly in his narratives. Summarising his narratives, the poem below indicates that both his religious faith and his mobilisation of local understandings of childhood have enabled Samuel to create a positive caregiver identity, based on the acceptance, love and blessings he gets in return for his caring, both from the community and God. Samuel therefore sees caregiving as a strategy through which he derives recognition and support from the community. In addition, Samuel distinguishes himself from other children, arguing that "if I have something I can share it with other children, I don't deny them." Samuel values this quality highly and attributes this good quality to his caregiving experiences. He believes that many other children "don't share what they have with others, they pretend not to have anything." The poem summarises and links the positive meanings that Samuel attaches to caregiving and the construction of a positive caregiving identity, one that influences Samuel's psychosocial well-being and coping.

I have a Helping Heart

How are you different?
I like helping people.

I have a helping heart,
if I have something I share it

What makes you happy?
All that I have done makes me happy
the villagers love me seriously
as I don't do bad things in the community

How is life as a career?
If I care for a sick person,
I'm happy since I get blessings from God
It is not good for one to suffer.
There are no negative effects

Carolyne, Age 15 (Mixed Meanings)

Carolyne's father died of an illness when she was 7 years old and she began taking care of her mother at the age of 10. For two years Carolyne provided nursing care and psychosocial support to her mother and kept the household running. She cooked and fed her, washed and massaged her body. In addition to the nursing care, "I also did the other house work such as cleaning the house, washing utensils and fetching water." The workload was heavy and interrupted with a change when her mother was admitted to the hospital. In the absence of adequate nursing care at the hospital, Carolyne continued to care for her mother in the hospital, forcing her to leave school. Carolyne's mother died in the hospital. During this time, she was largely coping by herself—possibly as a result of the stigma associated with AIDS. But her caring experiences do not end here.

Carolyne and her little sister then moved in with her grandmother who was very old and required care and support. All her children had died and she was dependent on her grandchildren. Despite taking on significant caring and household chores, Carolyne was determined to return to school. But soon after re-entering, her class teacher advised her to leave school as she was not attending classes consistently due to her responsibilities at home. In addition to her poor attendance, Carolyne suffered from slight visual impairment and had difficulties reading what was written on the blackboard. Determined to stay in school, Carolyne refused to drop out and continued with her education. Whilst dealing with her personal health issues and education, Carolyne also lost the support of her little sister: "My sister was sponsored [to go to school] by some whites and she is now in Nakuru." Aside from her sister, Carolyne does not mention receiving support in her provision of care from anyone else. This is also reflected by the amount of time she reported to spend on caring, "you work throughout the day, no time for resting." She did however occasionally negotiate her way to material or food support from community members. In describing a photo of a woman who has supported her, she says." This photo reminds me of the kind of support and love we get from the community members. If I need anything, I tell them and if it is available, I will get it."

After two years of caring, Carolyne's grandmother passed away. She was left to stay with her grandfather who was mistreating her, often leaving her to spend nights outside where she describes her encounters with hyenas. Carolyne quickly moved in with her aunt, who has been supporting her well since, "If I ask my aunt to buy me something she will not ignore me, even though I am not her child. No, she will do for me the same as for her own children." Carolyne and her aunt jointly provide care for her aunt's elderly co-wife.

Like Samuel, the numerous caring experiences that have confronted Carolyne have facilitated a caring identity. Unlike Samuel, her identity is a more reflective one, acknowledging both the negative and positive impact of caregiving. Although she is now in foster care with her aunt (whom she calls mother), she continues to provide care and support for the sick and old in her community, this time with help from her aunt. Although Carolyne says that "the lives of children caring for the sick is not good," she describes her circumstances and work as something she just had to do, without complaints, and something which is important. Carolyne's experience highlights both the difficulties she has faced in providing care, as well as the way she has dealt with them. Aside from the periodic support she has received from her sister and aunt, her social environment has been of limited support, yet she accepted her role as carer and simply 'got on' with the job.

As the poem summarising Carolyne's accounts demonstrate, Carolyne has mixed feelings toward her circumstances, acknowledging both the negative and positive impact caregiving has on her, consciously accepting her circumstances.

All that I have Done has been Important

> I was first caring for my mother
> when she was sick.
> She was too sick to do anything.
> I was the one to wash her and feed her.
> It reached a point where I could not sleep.

She was crying of pain all the time.

She needed water and wanted to be massaged.

I also did the housework.

Cleaned the house, fetched water and prepared food.

I had to leave school.

After her death my grandmother fell sick.

I started caring for her.

We had no money, I could not take her to the hospital

I have had problems.

I have been committed to caring and had little rest.

But all that I have done has been important.

Pascal, Age 14 (Negative Meaning)

Pascal began providing care from a very young age. Until his seventh birthday, Pascal was taking care of his father who suffered from AIDS. Pascal did not receive much support from his mother, who eventually left the house because of stigma, leaving 6-year-old Pascal alone with his father. Pascal's older siblings lived away and only provided limited support, mostly in terms of food. Pascal spoon fed his father and cleaned his body, also in the most intimate of places. When Pascal tried to seek out help from his mother who was staying with her brothers, he was chased away by his uncles. When asked about how he coped with the situation, Pascal explained: "I had a vegetable garden which I used to cultivate and sell the produce from in order to buy drugs for my father or anything else he needed." At the age of seven, Pascal's father died. Following his father's death, Pascal moved in with one of his brothers and returned to school. To Pascal's dismay, he was told by his teacher that he had to repeat class 2. Pascal had a difficult year: "It was painful to see my classmates in class 3. I did not forget the caring of my father for the whole first term and I could only think about how my father died. This thinking left me at the bottom of my class. But in term 2 I started to forget these things slowly." As with Pascal, many young carers attach a feeling of loss or damage to their education as a result of time consuming caregiving or a lack of concentration.

Pascal's caring experiences did not end here. A couple of years later his brother also got ill, but this time Pascal was not alone in caring: "Fortunately we were two of us, so one cared at night and another during the day. I especially cared during the night and it made me unable to concentrate as I almost slept in class." As his brother got bedridden, Pascal left school once again and had to repeat class four when his brother died. Following the death of his brother, Pascal moved in with his grandmother where he currently stays. He supports his grandmother with those tasks she cannot perform due to old age and is reviving his relationship with his mother, who decided to return to the community two years after his father's death. In reflecting on the limited support he received during his caregiving, which was a key contributor to the negative meanings he attach his experiences, Pascal laid heavy emphasis on the relative poverty that he endured: "Other children have their school fees paid for by their parents, they have good clothes, good shoes and they look nice whilst I cannot afford to look nice because of the little money I have." In comparing his life with other children, Pascal clearly sees himself as a victim and feels the injustice surrounding his circumstances. Despite these victimological representations, Pascal has not lost hope: "I know and hope that my life will be good."

The poem summarising Pascal's experience exemplifies the very difficult conditions in which Pascal was providing care. It is evident from Pascal's narratives that he feels a tremendous sense of loss and damage to his life, circumstances that have pushed Pascal to feel a sense of relief following the death of his father.

All this Suffering

This drawing (see figure 2) shows the kind of care I have given to the sick

My mother was nowhere to be found

My father's sickness got worse and worse

It forced me to leave school

He was unable to walk

I washed off his faeces

He disturbed me during the night

I was very sad, I was left alone with my father

When he died, I thanked God, he made me suffer a lot

One of my brothers fell sick, this also made me leave school

I had to repeat class four when my brother died

All this suffering made me go to my grandmother's place

I am now in class 7 and learning

I help my grandmother with harvesting,

fetching water and cooking

It shows good behaviour and the majority loves me.

I am still not happy with the kind of life I am living, though I am in school

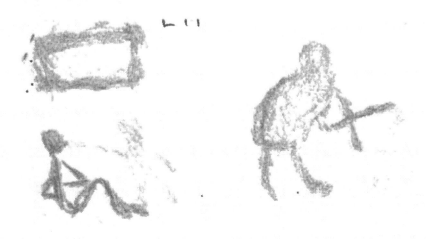

Figure 2. Drawing by Pascal.

Discussion

Our three case studies highlight the duties involved with chronic care of adults affected by HIV/AIDS. In the absence of working adults in their households, many of the children were responsible for the generation of food and income to sustain their household. Alongside these responsibilities, children showed great concern for the care and well-being of their sick parents or ageing grandparents, taking on considerable nursing duties. Children are often the primary caregivers for their ailing parents and play a fundamental role in the management of their parent's disease and opportunistic infections. Approaching young carers as social actors, the case studies illustrate how young carers cope with the care of guardians who suffered from chronic conditions. We have highlighted the social resources and protective factors potentially available to them (or not) and how these social resources may contribute to the way in which they assess their risk and vulnerabilities.

Samuel received support from his aunt and little sister and used his faith and local representations of childhood to construct a positive identity as a young caregiver. Samuel suggests he is loved by the community because he provides care and support, a reflection of the local expectation of childhood as a period of duty and service. Carolyne, on the other hand, did recognise the disadvantages and the problems she was facing, but also managed to attach some positive meanings to her experiences. She surrendered to her circumstances and got on with caring, approaching caregiving as a challenge, something difficult, but something which was important. Although Carolyne sees caregiving overall as positive, there were other children, able to reflect on a mix of positive and negative meanings, who perceived caregiving as an overall negative thing. This was often upon reflection on the impact caregiving has had on their education. We found only a few children with completely ambivalent meanings. Pascal was much less optimistic about his circumstances, reflecting on his particular situation and lack of social support. Pascal was further influenced by representations of the importance of education, and as caring compromised his education in certain periods, he felt a sense of loss and damage to his chances in life. As is evident from the case studies, their specific circumstances, social resources and representations impact on the meanings which the children attached to caring. This in turn contributes to their psychosocial well-being, a relationship (between ascribed meaning and well-being/coping) that has been documented in the context of physical illnesses [42,57,58].

Most of the children involved in the wider study attached a variety of meanings to caregiving and many were still negotiating the different meanings they attached to various contexts. These mixed meanings were often a reflection of the paradoxical expectations of children. On one hand the children draw on local understandings of childhood as a time of duty and service, and on the other, children are told about the importance of attending school. As young carers having to juggle with both duty and service and education, the meanings they ascribe to their circumstances are often a by-product of their negotiation between these expectations. Nevertheless, the negotiation of meanings that these children actively engage in facilitates their psychosocial coping and connects them to different meanings at different times in order to achieve particular goals.

Whilst certain caregiving circumstances may well be detrimental to a child's mental health, this is not always the case. Counter balancing (or not) some of their vulnerabilities were the symbolic and socioeconomic resources that arose from their familial and social environments, allowing many of the children to draw on some of the more positive aspects of caregiving. The ability of caregiving children to identify and draw on the benefits of their difficult circumstances has previously been identified as a strategy for psychological survival [19]. The three case studies presented in this paper also suggest that the children's previous experiences of caregiving and whom they currently care for is an important influence on the meanings they attach. Carolyne for example, who has provided care and support to a number of adults over a period of time, is now used to caregiving and has a break from the emotional cost of caring for an ailing parent. Carolyne is able to reflect on the hard times she had, but also on the skills and social benefits caregiving has brought her. Carolyne was not an exception. Many of the participating young carers have had numerous caregiving experiences and continue to provide care long after parental death, as their grandparents, who have lost children to AIDS, are in need of care and support as they age. As exemplified by Carolyne, the majority of children, depending on the context and time, approach and describe their circumstances differently, and effectively move around within the continuum of psychosocial coping. Some children have actively constructed a positive caregiving identify, whilst others see themselves as marginalised. These meanings and identities may change with time. These findings suggest that the vulnerability of young carers is a process [59], one which requires a nuanced understanding of the protective factors and social resources available for the children to actively draw [54].

One important finding is that the children are able to identify numerous benefits to young caregiving and actively draw on these benefits to facilitate psychosocial coping. We hope that our analysis has shown that to provide appropriate support for young caregivers, health professionals must align their understanding and responses to the psychosocial cost of chronic care, to a more nuanced and contextual understanding of social agency and opportunities evident in African communities. Rather than focusing on counselling services, interventions targeting young carers should strengthen the existing social resources within their context and provide the communities with the financial and social psychological resources to do so.

Competing Interests

The authors declare that they have no competing interests.

Authors' Contributions

The study was conceived and coordinated by MS. Both authors contributed to the conception and design of the study. VO participated in conducting the interviews and workshops from which the data were collected as well as transcribing and translating the data. MS drafted the paper. Both authors read and approved the final manuscript.

Acknowledgements

Our special thanks go to all the children in this study who enthusiastically took part in this project. We would also like to thank the reviewers of this paper for their constructive comments and Prof. Cathy Campbell from the LSE Health, Community and Development Research Group for helpful comments on an earlier version of this paper. Lastly, we would like to thank Cellestine Aoro for her role in the data collection and preparation.

References

1. Prins G: AIDS and global security. International Affairs 2004, 80:931–952.

2. Beaudin CL, Chambre SM: HIV/AIDS as a Chronic Disease: Emergence From the Plague Model. American Behavioral Scientist 1996, 39:684–706.

3. Thorne S: Negotiating health care: the social context of chronic illness. Newbury Park, CA: Sage; 1994.

4. UNAIDS: 2008 Report on the global AIDS epidemic. [http:/ / www.unaids. org/ en/ KnowledgeCentre/ HIVData/ GlobalReport/ 2008/ 2008_Global_report.asp] Geneva: UNAIDS; 2008. accessed 15/01/2009

5. UNAIDS/WHO: 2007 AIDS Epidemic Update. Geneva, Switzerland: UNAIDS/WHO; 2007.

6. Russel S, Seeley J, Ezati E, Wamai N, Were W, Bunnell R: Coming back from the dead: living with HIV as a chronic condition in rural Africa. Health Policy and Planning 2007, 22:344–347.

7. Hardon AP, Akurut D, Comoro C, Ekezie C, Irunde HF, Gerrits T, Kglatwane J, Kinsman J, Kwasa R, Maridadi J, et al.: Hunger, waiting time and transport costs: Time to confront challenges to ART adherence in Africa. AIDS Care 2007, 19:658–665.

8. Kipp W, Tindyebwa D, Rubaale T, Karamagi E, Bajenja E: Family Caregivers in Rural Uganda: The Hidden Reality. Health Care for Women International 2007, 28:856–871.

9. Unge C, Johansson A, Zachariah R, Some D, Van Engelgem I, Ekstrom AM: Reasons for unsatisfactory acceptance of antiretroviral treatment in the urban Kibera slum, Kenya. AIDS Care 2008, 20:146–149.

10. Dahab M, Charalambous S, Hamilton R, Fielding K, Kielmann K, Churchyard G, Grant A: "That is why I stopped the ART": patients & providers' perspectives on barriers to end enablers of HIV treatment adherence in a South African workplace programme. BMC Public Health 2008, 8:1–6.

11. Posse M, Meheus F, van Asten H, Ven A, Baltussen R: Barriers to access to antiretroviral treatment in developing countries: a review. Tropical medicine & international health 2008, 13:904–913.

12. Mshana GH, Wamoyi J, Busza J, Zaba B, Changalucha J, Kaluvya S, Urassa M: Barriers to accessing antiretroviral therapy in Kisesa, Tanzania: A qualitative study of early rural referrals to the national program. Aids Patient Care and Stds 2006, 20:649–657.

13. Chen S, Yu J, Harries A, Bong C, Kolola-Dzimadzi R, Tok T, King C, Wang J: Increased mortality of male adults with AIDS related to poor compliance to antiretroviral therapy in Malawi. Tropical medicine & international health 2008, 13:513–519.

14. Chimwaza A, Watkins S: Giving care to people with symptoms of AIDS in rural sub-Saharan Africa. AIDS Care 2004, 16:795–807.

15. Kipp W, Tindyebwa D, Karamagi E, Rubaale T: How much should we expect? Family caregiving of AIDS patients in rural Uganda. Journal of Transcultural Nursing 2007, 18:358–365.

16. Shifren K: How Caregiving Affects Development: Psychological Implications for Child, Adolescent and Adult Caregivers. Washington, DC: American Psychological Association; 2009.

17. Nkosi T, Kipp W, Laing R, Mill J: Family caregiving for AIDS patients in the democratic republic of Congo. World Health & Population 2006, 8:4–13.

18. Robson E, Ansell N: Young carers in Southern Africa: Exploring stories from Zimbabwean secondary school students. In Children's Geograpahies: Playing, Living, Learning. Edited by: Holloway S, Valentine G. London: Routledge; 2000:174–193.

19. Robson E, Ansell N, Huber U, Gould W, van Blerk L: Young Caregivers in the Context of the HIV/AIDS pandemic in sub-Saharan Africa. Population, Space and Place 2006, 12:93–111.

20. Robson E: Invisible carers: Young people in Zimbabwe's home-based health care. Area 2000, 32:59–69.

21. Robson E: Hidden Child Workers: Young Carers in Zimbabwe. Antipode 2004, 36:227–248.

22. Robson E: Interviews Worth the Tears? Exploring the Dilemmas of Research with Young Carers in Zimbabwe. Ethics, Place and Environment 2001, 4:135–142.

23. Evans R, Becker S: Children caring for parents with HIV and AIDS: Global issues and policy responses. Bristol: Policy Press; 2009.

24. Metzing S: Kinder und Jugendliche als pflegende Angehörige: Wie sich pflegerische Hilfen auf ihr Leben auswirken können. Eine internationale Literaturstudie (1990–2006). Pflege 2007, 20:331–336.

25. Olsen R: Young carers: challenging the facts and politics of research into children and caring. Disability and Society 1996, 11:41–54.

26. Cluver L, Gardner F, Operario D: Psychological distress amongst AIDS-orphaned children in urban South Africa. Journal of Child Psychology and Psychiatry 2007, 48:755–763.

27. Makame V, Ani C, Grantham-McGregor S: Psychological well-being of orphans in Dar El Salaam, Tanzania. Acta Paediatr 2002, 91:459–465.

28. Ruiz-Casares M, Thombs B, Rousseau C: The association of single and double orphanhood with symptoms of depression among children and adolescents in Namibia. European Child and Adolescent Psychiatry 2009, 18:369–376.

29. Bhargava A: AIDS epidemic and the psychological well-being and school participation of Ethiopian orphans. Psychology, Health and Medicine 2005, 10:263–275.

30. Lester P, Rotheram-Borus M, Lee S, Comulada S, Cantwell S, Wu N, Lin Y: Rates and predictors of anxieety and depressive disorders in adolescents of parents with HIV. Vulnerable Children and Youth Studies 2006, 1:81–101.

31. Cluver L, Gardner F: The Psychological Well-being of children orphaned by AIDS in Cape Town, South Africa. Annals of General Psychiatry 2006, 5:8.

32. Atwine B: Psychological distress among AIDS orphans in rural Uganda. Social Science & Medicine 2005, 61:555–564.

33. Cluver L, Gardner F: The Mental health of children orphaned by AIDS: a review of international and southern African research. Journal of child and adolescent Mental Health 2007, 19:1–17.

34. Bauman L, Foster G, Silver E, Berman R, Gamble I, Muchaneta L: Children caring for their ill parents with HIV/AIDS. Vulnerable Children and Youth Studies 2006, 1:56–70.

35. Bauman L, Johnson Silver E, Berman R, Gamble I: Children as caregivers to their ill parents with AIDS. In How Caregiving Affects Development: Psychological Implications for Child, Adolescent and Adult Caregivers. Edited by: Shifren K. Washington: American Psychological Association; 2009:37–63.

36. Cluver L: Young Carers South Africa Project. [http://www.youngcarers.netau.net] Department of Social Policy and Social Work, University of Oxford; 2009.

37. Zhang L, Li X, Kaljee L, Fang X, Lin X, Zhao G, Zhao J, Hong Y: I felt I have grown up as an adult: caregiving experience of children affected by HIV/AIDS in China. Child: Care, Health and Development 2009, 35:542–550.

38. Cree V: Worries and problems of young carers: issues for mental health. Child and Family Social Work 2003, 8:301–309.

39. Boris N, Brown L, Thurman T, Rice J, Snider L, Ntaganira J, Nyirazinyoyo L: Depressive symptoms in youth heads of household in Rwanda – Correlates and implications for intervention. Archieves of pediatrics and adolescent medicine 2008, 162:836–843.

40. Antonovsky A: Health, Stress and Coping. San Francisco: Jossey-Bass; 1979.

41. Antonovsky A: Unravelling the Mystery of Health – How People Manage Stress and Stay Well. London: Jossey-Bass Publications; 1987.

42. Luker K, Beaver K, Leinster S, Owens R: Meaning of Illness for women with breast cancer. Journal of Advanced Nursing 1996, 23:1194–1201.

43. Farber E, Mirsalimi H, Williams K, McDaniel J: Meaning of Illness and Psychological Adjustment to HIV/AIDS. Psychosomatics 2003, 44:485–491.

44. GOK: Bondo District Monitoring and Evaluation Report 2003–2004. Government of Kenya: Ministry of Planning and National Development: Accessbile from Bondo District Resource Centre; 2005.

45. GOK: Bondo District Development Plan 2002–2008. Government of Kenya: Ministry of Finance and Planning: Accessible from Bondo District Resource Centre; 2002.

46. NACC: Kenya HIV/AIDS Data Booklet 2005. Republic of Kenya: National AIDS Control Council; 2005.

47. UNAIDS/WHO: Epidemiological Fact Sheet on HIV and AIDS – Core data on epidemiology and response Kenya. [http://www.who.int/globalatlas/predefinedReports/EFS2008/full/EFS2008_KE.pdf]. Geneva: Switzerland: UN-AIDS/WHO Working Group on Global HIV/AIDS and STI Surveillance; 2008. accessed 28/11/2008

48. Nyambedha E, Wandibba S, Aagaard-Hansen J: Changing patterns of orphan care due to the HIV epidemic in Western Kenya. Social Science & Medicine 2003, 57:301–311.

49. Roberts H: Listening to Children: and Hearing Them. In Research with Children: Perspectives and Practices. Edited by: Christensen P, James A. New York: Routledge; 2008:260–275.

50. O'Kane C: The development of participatory techniques: facilitating children's views about decisions which affect them. In Research with Children: Perspectives and Practices. Second edition. Edited by: Christensen P, James A. New York: Routledge; 2008:125–155.

51. Wang C: Youth Participation in Photovoice as a Strategy for Community Change. In Youth Participation and community change. Edited by: Checkoway B, Gutiérrez LM. Binghamton, NY: Haworth Press, Inc; 2006:147–161.

52. Wang C, Burris M: Photovoice: Concept, Methodology, and Use for Participatory Needs Assessment. Health Education & Behaviour 1997, 24:369–387.

53. Wang C, Yi W, Tao Z, Carovano K: Photovoice as a participatory health promotion strategy. Health Promotion International 1998, 13:75–86.

54. Skovdal M, Ogutu V, Aoro C, Campbell C: Young Carers as Social Actors: Coping Strategies of Children Caring for Ailing or Ageing Guardians in Western Kenya. Social Science and Medicine 2009, 69:587–595.

55. Gee J: A Linguistic Approach to Narrative. Journal of Narrative and Life History 1991, 1:15–39.

56. Poindexter C: Meaning from Methods: Re-presenting Narratives of an HIV/affected Caregiver. Qualitative Social Work 2002, 1:59–78.

57. Lipowski Z: Physical illness, the individual and the coping processes. Psychiatry and Medicine 1970, 1:91–102.

58. Barkwell D: Ascribed Meaning: A Critical Factor in Coping and Pain Attenuation in Patients with Cancer-Related Pain. Journal of Palliative Care 1991, 7:5–14.

59. Blum R, McNeely C, Nonnemaker J: Vulnerability, Risk and Protection. Journal of Adolescent Health 2002, 31:28–39.

Evolution of Patients' Complaints in a French University Hospital: Is there a Contribution of a Law Regarding Patients' Rights?

Camila Giugliani, Nathalie Gault, Valia Fares, Jérémie Jegu, Sergio Eleni dit Trolli, Julie Biga and Gwenaelle Vidal-Trecan

ABSTRACT

Background

Legislative measures have been identified as one effective way of changing attitude or behaviour towards health care. The aim of this study was to describe trends in patients' complaints for medical issues; to evaluate the contribution of a law regarding patients' rights, and to identify factors associated to patients' perception of a medical error.

Methods

Patients with a complaint letter for medical issues in a French university hospital were included. Trends in complaint rates were analysed. Comparisons were made between a first (1998–2000) and a second (2001–2004) time period, before and after the diffusion of the law, and according to the perception of a medical error.

Results

Complaints for medical issues increased from 1998 to 2004. Of 164 complaints analysed, 66% were motivated by the perception of a medical error (47% during the first time period vs. 73% during the second time period; p = 0.001). Error or delay in diagnosis/treatment and surgical/medical complication were the main reasons for complaints. Surgical departments had the higher number of complaints. Second time period, substandard care, disability, and adverse effect of a health product were independently associated with the perception of a medical error, positively for the formers, and negatively for the latter.

Conclusion

This study revealed an increase with time in the number of complaints for medical issues in a university hospital, as well as an increase in the perception of a medical error after the passing of a law regarding patients' rights in France.

Background

Patient safety and risks of inpatient care are current issues of concern worldwide and should be a declared priority within health care organizations, since adverse events account for unacceptable high levels of patient morbidity and mortality [1-3], varying from 3,7% in France or in the New-York State [4-6] to around 10% in Great-Britain or New-Zealand [5-8]. Many countries have guaranteed patients' right to process for resolving dissatisfactions with health care providers. The United States, for example, since November 1997 have included an aspirational statement in the Consumer Bill of Rights regarding this issue [9]. In this country, where iatrogenesis is the third leading cause of death [10], increasing patients' participation in their care, reducing health care errors, and ensuring the appropriate use of health care services are among the national aims for improving the quality of health care [11,12]. Other countries, such as Sweden, the United Kingdom, Italy, Spain, and New Zealand also have a specific legislation regarding patient's protection and safety [13,14]. In France, considerable efforts have been made in the past 20 years to improve in-hospital safety management through laws and regulations [15-17]. Furthermore, patients found their rights supported by a law [17] developed in 2001 by the health authorities and voted in by Parliament on 4 March 2002. This law concerned among others: the respect of dignity and the absence of discrimination regarding patients' care, better identification and management of the consequences associated with poor sanitation (right for compensation for no-fault medical accidents and better access to medical insurance for persons with a serious health risk), and assurance of direct access to medical records. Moreover, a national institution depending on the Ministry of Health was created (National Organism of Financial Compensation of Medical Accidents, iatrogenesis and nosocomial infections, the ONIAM [Office National d'Indemnisation des Accidents Médicaux, des affections iatrogènes et des infections nosocomiales]. This institution is in charge of the organisation of the plan for out-of-court resolution of medical accidents and of financial compensation of patients in case of no-fault medical accident. In addition, the creation of a commission of relations with patients became a requirement to every hospital. All these measures may have contributed to an increased expectation of patients towards the health system and may have facilitated the act of officially addressing a complaint.

A complaint can express grief or resentment, resulting in some cases in financial compensation. However, some studies have found that the main reason for patients to file a complaint was not the desire for financial compensation but to expose that things went wrong, and to get an answer from the hospital or doctors stating that the reoccurrence of the situation will be prevented [18,19].

Legislative measures have been identified as one effective way of perpetuating a change in attitude or behaviour regarding health care [20]. In terms of patients' rights within the health system, we hypothesized that the law has facilitated the complaining process. However, published data regarding the effects of legislative measures on patients' complaints is limited. Thus, this study reports an analysis of patients' complaints regarding medical issues, from the patient's perspective, in a university hospital during a period of 6.3 years (1998 through 2004), including (1) a description of their trends, (2) an identification of factors associated with patients' perception of a medical error, and (3) an evaluation of the contribution of the law voted on March, 4th 2002.

Methods

Study Design

We conducted an analysis of patients' complaint letters received by the Direction of a university hospital from 01/01/1998 to 30/03/2004, considering the patient's perspective.

Setting

The study was performed in a hospital belonging to the Assistance Publique – Hôpitaux de Paris, the Public Network of Parisian Hospitals (41 hospitals). This hospital complex (1,316 short-stay beds at two different locations) houses all major medical and surgical departments, except nephrology, neurosurgery, and cardiovascular surgery. In 1999, a paediatric hospital joined the hospital complex. A medical department of public health is in charge of quality and safety of health care. The hospital's Direction includes a Department of Patient's Rights which is responsible for handling all written complaint letters coming from four main sources: the litigation department of the Public Network of Parisian Hospitals; the court, the President's administration of justice, which transmits complaints to the litigation department of the Public Network of Parisian Hospitals; the hospital's clinical departments, especially the Emergency department; and the Department of Patient's Rights itself, which also receives complaints written by all the previous sources.

The department of Patient's Rights investigates every complaint and tries to solve it out of court. In case of failure, the case is transmitted to the litigation department of the Public Network of Parisian Hospitals. In case of no-fault medical accident, patients can refer their case to a regional commission of conciliation, which decides the allocation of a financial compensation without litigation [17].

Study Population

Every inpatient with a written complaint regarding medical issues, with or without legal involvement, coming in the hospital's Department of Patient's Rights within the study period was included. Both inpatients and outpatients having visited the Emergency Department were considered eligible. Complaints exclusively for financial issues (e.g., hospitalisation fees), accommodation, hosting quality, and organisational matters, were excluded from the sample by the staff of the Department of Patient's Rights. Complaints that were not excluded for these criteria were considered as complaints for medical issues.

Data Collection

The number of hospital admissions (excluding day-care) and of visits to the Emergency Department was drawn from the annual hospital activity reports. Within the study period, the hospital registered 705,632 both outpatient visits to the Emergency Department and inpatient admissions in other clinical departments. The Department of Patient's Rights received a total of 2,116 complaint letters, of which 164 were for medical issues, selected by the Director of quality and patients' rights and one of the investigators (VF). We chose to collect only the complaints for medical issues because these are more related to health providers' practice, consequently, medical errors, which, amongst the reasons of complaints, is the main focus of this study. Information contained in the complaint letters for medical issues was collected using an anonymous semi-structured form filled in by the same investigator from the medical department of public health. This form was conceived according to the content analysis of the complaints. Data collected included patient's demographics, type of issue, type of department, patient's perception of a medical error, type of patient's outcome after the perceived medical issue and demand for financial compensation. The medical department of public health was in charge of coding the written complaints, and handled the raw data. Complaints were classified by confronting blind conclusions of three investigators from the medical department of public health (GVT, SET et JB) for three main issues: the perception of a medical error, the reason of complaint and the consequences of the accident leading to the written complaint. In case of disagreement between them, a case discussion took place to reach a consensus opinion.

Definition of Variables

The complaint rates were calculated by dividing, first, the number of complaints for medical issues by the total number of complaints received by the Department of Patient's Rights, and second, the number of complaints for medical issues by the hospital activity.

We defined two different time periods, considering the large amount of media coverage related to the passing of the law in 2001, previously to its implementation: 1998–2000 ("first period") and 2001–2004 ("second period"). Another definition could have been considered (1998–2001 and 2002–2004, for instance); but the "first period" was chosen after a sensitivity analysis of the results for two time periods (i.e., 1998–2000 and 1998–2001). The comparison of complaint's characteristics between 1998–2001 and 2002–2004 lead to non significant differences. The results of our analysis were sensitive to the changes in thresholds of time periods.

Content analysis of the letters allowed us to classify reasons for complaints according to patient's perspective in two different variables:

(1) The types of issue were listed as error or delay in diagnosis or in treatment (i.e. wrong or delayed diagnosis or medication, resulting in adverse outcome), nosocomial infection, the adverse effects of a health product (i.e., blood product, drug, graft, medical device), results of a surgical or a medical complication, information or surveillance problem (i.e. lack of interest, attention or care), non performance of a required procedure or

treatment and other. Error or delay in diagnosis or in treatment, and information or surveillance problem were considered as parts of a "substandard care" category (i.e. procedures not performed according to standard of care). Results of a surgical or a medical complication and nosocomial infection were grouped together in a "complication of care" category.

(2) The outcomes were listed as death, disability, surgery (operation or re-operation), prolonged care (e.g. hospital stay), failure to achieve standard of care, and other. Surgery, prolonged care and failure to achieve standard of care were considered as parts of a "prolonged care" category.

We defined two types of departments, those functioning essentially on an emergency basis (adult and paediatric emergency departments, obstetrics, gynaecology, and neonatology), and the others.

Statistical Analysis

First, trends in complaint rates for medical issues were examined on a yearly basis (number of complaints among hospital activity and complaints for medical issues among all complaints). The association between the rates of complaints and years was measured using Pearson correlation coefficient. We further analysed the evolution of the number of complaints according to time (years) and the existence of the law, using a Poisson regression, however, the Variance Inflation Factor (VIF) and tolerance exploring the co linearity between these variables led us not to use the results of this model.

Second, we performed a descriptive study of complaint letters, using proportions and their 95 percent confidence intervals (95% CI). Third, we compared the characteristics of the written complaint between the first and the second periods, and according to the perception of a medical error, using chi-square tests.

We further explored the association of these differences with the perception of a medical error using a logistic regression model. Variables with $p < 0.05$ were included in the model, run with a conditional backward stepwise procedure. The associations were expressed as odds ratios (ORs) with their 95% CI. Diagnosis of the regression model and robustness were checked.

All tests were two-tailed, and statistical significance was set at $p < 0.05$. The software SPSS® for Windows™, version 12.0 (Copyright© SPSS Inc., 2003) was used for data analysis.

Ethical Considerations

This was a study of patients' complaints, with anonymous data collection, thus demanding no specific ethics approval.

Results

Trends

The rate of complaints for medical issues according to hospital activity increased yearly (Table 1) with a correlation coefficient of Pearson of 0.87 (p = 0.01). The rate of complaints for medical issues among all complaints also increased, from 5.4% in 1998 to 14.2% in 2004 (Table 1), with a coefficient of Pearson of 0.95 (p = 0.001). The evolutions of complaints are illustrated in a clear trend (Figure 1).

Table 1. Rates of complaints in a university hospital of the Public Network of Parisian Hospitals from 1998 to 2004

Year	Complaints for medical issues among all complaints n/N (%)	Complaints for medical issues among hospital activity n/N* (%)
1998	11/202 (5.4)	11/97 417 (0.011)
1999	11/254 (4.3)	11/125 756 (0.009)
2000	25/463 (5.4)	25/106 688 (0.023)
2001	31/417 (7.5)	31/115 983 (0.027)
2002	37/408 (9.2)	37/115 618 (0.032)
2003	32/272 (11.9)	32/116 619 (0.027)
2004†	17/123 (14.2)	17/27 551 (0.062)

* N = emergency visits and inpatients, excluding one-day hospitalizations
† Only the period of January through March has been studied

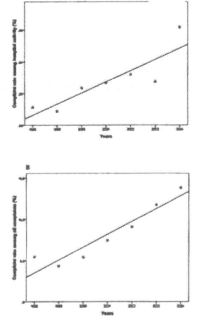

Figure 1. Evolution of complaints for medical issues among hospital activity (a) and complaints for medical issues among all complaints (b) between 1998 and March 2004.

Characteristics of Study Population and Complaints

Women accounted for 63% (95% CI = 55% – 70%) of complaint letters for medical issues. Median age was 39 years (range: 0 – 88). Patients less than 15 accounted for 22% (95% CI = 16% – 30%) of letters. The complaint was written by a lawyer in 42% of cases (95% CI = 35% – 50%), whereas the affected patient was the author in 41% of cases (95% CI = 33% – 49%). Seventeen percent of letters were written by relatives (95% CI = 11% – 23%). Eleven complaints (6.7%, 95% CI = 3.4% – 11.7%) were referred to the regional commission of conciliation. Among them, only one was further referred to a court.

Complaint letters were mostly due to error or delay in diagnosis or treatment (n = 40, 24%, 95% CI = 18% – 32%), results of a surgical or a medical complication (n = 40, 24%, 95% CI = 18% – 32%), or adverse effects of a health product (n = 27, 17%, 95% CI = 11% – 23%). The other issues reported were the non performance of a required procedure or treatment or other (n = 23, 14%, 95% CI = 9% – 20%), information or surveillance problem (n = 18, 11%, 95% CI = 7% – 17%), and nosocomial infection (n = 16, 10%, 95% CI = 6% – 15%). In 108 patients, (66%, 95% CI = 58% – 73%), the complaint was perceived as a medical error. Complaints for medical issues were more frequent in the surgical and obstetrical departments (respectively 39% and 23%), so as the medical error perceived by the patient (respectively 30% and 28%). Death was reported in 26 complaint letters (16%, 95% CI = 11% – 22%). For 45 patients (27%, 95% CI = 21% – 35%), disability was the final outcome. The other reported outcomes were prolonged care (n = 11, 7%, 95% CI = 3% – 12%), surgery (n = 10, 6%, 95% CI = 3% – 11%), failure to achieve standard of care (n = 9, 5%, 95% CI = 3% – 10%) and others or unknown (n = 63, 38%, 95% CI = 31% – 46%).

Comparisons of the Characteristics between the Two Periods (Table 2)

Complaints regarding younger patients, departments functioning essentially on an emergency basis and perception of a medical error were significantly more frequent in the second period.

Table 2. Comparison of complaints for medical issues between two time periods (1998–2000 and 2001–2004) in a university hospital of the Public Network of Parisian Hospitals*

Variable	Time period		
	First n/N (%)	Second n/N (%)	p
Age > 39	25/40 (63)	40/95 (42)	0.030
Male gender	19/47 (40)	42/117 (36)	0.390
Departments functioning on an emergency basis†	12/47 (26)	52/117 (44)	0.025
Perception of a medical error by the patient	22/47 (47)	86/117 (73)	0.001
Reason of complaint			0.119
Invalidity in diagnosis or treatment	8/42 (17)	32/117 (27)	
Nosocomial infection	6/42 (13)	10/117 (9)	
Adverse effect of a health product	10/42 (21)	17/117 (15)	
Result of a surgical/medical procedure	16/42 (34)	24/117 (21)	
Information or surveillance problem	3/42 (4)	16/117 (14)	
Non performance of a required procedure or treatment, or others	5/42 (11)	18/117 (15)	
Consequences			0.494
Death	9/47 (19)	17/117 (15)	
Disability	10/47 (21)	35/117 (30)	
Prolonged care	7/47 (15)	23/117 (20)	
Other or unknown	21/47 (45)	42/117 (36)	

* N is according to each variable, depending on non-missing values.
†Obstetrics, gynecology, neonatology and Emergency Departments.

Comparisons of Patient's Characteristics According to the Perception of a Medical Error (Table 3)

Complaints regarding departments functioning essentially on an emergency basis, and made in the second period were significantly more often associated with the perception of a medical error. Complaints due to substandard care and complaints where disability was the main outcome were significantly associated with the perception of a medical error, whereas complaints due to the adverse effect of a health product were associated with the non-perception of a medical error. After adjustment, the second period, substandard care and disability as the main outcome were positively associated with the perception of a medical error, whereas complaints due to the adverse effect of a health product were negatively associated.

Table 3. Characteristics of patients, hospitalization and complaints associated with the perception of a medical error in a university hospital of the Public Network of Parisian Hospitals (conditional backward stepwise logistic regression model)

Variable	Perception of a medical error			Crude OR* (95% CI)	Adjusted OR (95% CI)
	Yes n/N (%)	No n/N (%)	p		
Age > 39	39/87 (45)	26/48 (54)	0.299		
Male gender	36/108 (33)	25/56 (46)	0.153	0.6 (0.3 – 1.2)	-
Departments functioning on an emergency basis ‡	50/108 (46)	14/56 (25)	0.008	2.6 (1.3 – 5.3)	-
Financial compensation	15/64 (23)	3/26 (8)	0.136		
Second period (2001–2004)	86/108 (80)	31/56 (55)	0.001	3.2 (1.6 – 6.4)	2.4 (1.0 – 5.3)
Type of issue			<0.001		
Substandard care §	48/108 (44)	10/56 (18)		2.4 (1.3 – 4.3)	2.2 (1.1 – 4.3)
Complication of care ¶	31/108 (29)	24/56 (43)		0.6 (0.4 – 1.1)	0.6 (0.4 – 1.2)
The adverse effect of a health product	10/108 (9)	18/56 (32)		0.3 (0.1 – 0.8)	0.3 (0.3 – 0.7)
Non-performance of a required procedure or treatment, other	19/108 (18)	4/56 (7)		-	-
Outcomes			0.004		
Death	17/108 (16)	9/56 (16)		0.85 (0.4 – 1.7)	0.7 (0.3 – 1.5)
Disability	38/108 (35)	7/56 (13)		2.4 (1.3 – 4.8)	2.4 (1.1 – 4.8)
Prolonged care	21/108 (19)	9/56 (16)		1.0 (0.5 – 2.0)	1.0 (0.5 – 2.1)
Other, unknown	32/108 (30)	31/56 (55)		-	-

* Odds Ratios
† 95% Confidence Interval
‡ Obstetrics, gynecology, neonatology and Emergency Departments
§ Invalidity in diagnosis or treatment, Information or surveillance problem
¶ Results of a surgical or a medical complication, nosocomial infection

Discussion

In this study based on the analysis of written complaints to the Department of Patient's Rights of a French university hospital, we found a time trend in the increments of the rate of complaints for medical issues related to the hospital activity and in the proportion of complaints for medical issues among all complaints. On the one hand, we observed that complaints referring to substandard care or complaints from patients whose outcomes had resulted in disability were positively associated with the perception of a medical error and this change appears to have been associated to the release of a law regarding patients' rights and quality of the health care (implemented in 2002). On the other hand, complaints referring to the adverse effect of a health product were negatively associated with the perception of a medical error. Moreover, complaints for medical errors were more frequent in the second period (2001 – 2004).

To our knowledge, this study is among the first analyses of patients' complaints in a university hospital in France. The examination of patients' complaints has been used by others [21] as patient's safety indicators in health care facilities. In other countries, some studies have concluded that the number of complaints increased [8,22-24]. In our study, an incremental impact of the law on the overall number of complaints could not be detected; the increment with time that we found might be a trend over a long period. In fact, one study, in the United Kingdom [23], found an increase of complaints after the implementation of a new complaint procedure.

Several reasons (e.g., improvement in organisational aspects due to engagement in accreditation procedures, reduction in length of hospital stay) could have led to a decrease in the number of complaints, especially complaints for medical issues. Nevertheless, we found an increasing number of complaints for medical issues despite organizational improvement, which suggests that this kind of complaint may not depend on organizational aspects. Other contributors to the increase in the proportion of complaints for medical issues among all complaints might be the presence of less inappropriate complaints for all matters, and a greater search for guilt within complaints. Other causes should still be considered, such as patients' greater consciousness of their rights and broader public debate about adverse events and health care errors, leading to a more demanding attitude towards the health care system (e.g. demand for financial compensation). The law may have lead the complainants to identify complaints as caused by a medical error perhaps because people think that a complaint for a medical error is more likely to lead to a financial compensation, even if there are other important reasons motivating people to complain, as mentioned above. We should keep in mind that a multivariate regression could have provided more accurate results, in particular, the independent effect of the time and that of the law. However, the small number of complaints each year, and the colinearity between the time variable and the law variable, led to misleading results that cannot be interpreted. In our study, we observed an increase of complaints written by lawyers; the reason could be a better and facilitated access to medical records and the current trend in France to increasing lawsuits for medical errors.

Complainants in this study were mainly middle-aged women. Such characteristics are supported by other publications [18,19,25,26]. Moreover, our findings confirmed the results found by other authors [6,8,27], in that the departments mostly affected by patients' complaints were surgery, obstetrics, gynaecology and neonatology.

We observed that reasons for complaints were mainly a result of a surgical or a medical complication, such as in an Australian study [28], and "substandard care" (error or delay in diagnosis or treatment, and information or surveillance problem). One American study [27] of the causes of adverse events in 1,047 inpatients reported the adverse effects following a particular procedure as one of the most common causes. Other authors have also reported concerns with standard of care among the main reasons for complaints [22,29], reflecting increased awareness of desirable standards of quality of care, or indeed patients' high expectations. Two studies [19,22] reported information problems within the three main reasons for complaint. Regarding the adverse effects of a health product, the results of a national study in French health facilities also reported this as the second most common cause of serious adverse events [6]. We found no other study reporting this particular reason of complaint. One possible explanation is that this could be accounted apart from complaints regarding hospital care directly, belonging, thus, to other statistics on adverse events.

In our study, a significant number of complaints were associated with death or disability (16% and 27% respectively). A recent French national study [6] reported 8% and 22% of death and disability, respectively, among the outcomes of serious adverse events, not far from the figures found in some British and American hospitals [4,8]. The higher rate of death found in our study could be explained by the adopted perspective (patients or relatives); actually, death is an outcome more traumatic than others and, thus, possibly more often the object of a complaint, even if it is not the result of an adverse event.

The impact of complaints can be diverse. Whilst complaints can be intended to promote better quality of health care and safer practices, two studies from New Zealand have shown the negative impacts of patients' complaints on surgical trainees training and performance (e.g. more defensive practice, less enjoyment and more stress) [25], and on surgeons practice (increased defensiveness and not learning from complaints) [26]. These studies have also shown that there is usually no appropriate support or guidance to deal with the harmful effects of complaints on medical practice, and that organisational support is needed, as well as an environment that encourages open disclosure and learning from mistakes.

In our study, the majority (66%) of complaints for medical issues was viewed by the patient as medical errors, whether this perception was founded or not. Such a perspective is not rare. In a study of public opinion in the United States [30], a national telephone survey performed in 1997, 42% of respondents reported that they or their close friends and relatives had already experienced a medical mistake. Moreover, the professional and the patient may have different opinions regarding the incident; what a patient believes to be an error may not have been perceived as such by the professional, or may have been caused by a chain of events [19]; usually this is a very complex matter.

In terms of the time period, we found that the perception of a medical error by the patient was more frequent on the second period of our study, which might reflect a change in the attitude of patients after the implementation of a legislative measure. We observed that patients were younger in the second period (2002–2004), a fact that can be explained by the addition of a paediatric hospital to the medical centre in 1999. However, this difference regarding the patient population had no influence in the perception of a medical error. The association that we observed between complaints for substandard care and perception of a medical error could be explained by an increase in the population's awareness of desirable standards of care.

On the one hand, we noticed that complaining for disability was more often associated with the perception of a medical error. This could be explained by the fact that, in this case, patients are more demanding for a financial compensation, as an amend to a functional loss. On the other hand, death was not associated with the perception of a medical error, probably because in this case, families are more often demanding for explanations and apologies.

Complaints are frequently considered the most acceptable option for expressing frustration and disappointment with the health care provided, but they might also be intended to promote better quality of care, in the sense of minimizing preventable health care related injuries. Moreover, in case of litigation, complaints can lead to economic sanctions on those who provide substandard care that leads to injuries.

Our study has some limitations. Since we found no typology for reason of complaints, we adapted typologies both from our analyses of complaint letters and from other studies [7,8,28]. Moreover, the classification method with three investigators could have led to classification bias. Indeed, our study intended to contain a qualitative analysis of written complaints. Since, in case of disagreement, a consensus opinion between investigators was reached, no interrater agreement was calculated. Other limitations were the small size of the sample, leading to a possible lack of power for detecting factors associated with the observed findings, retrospective design of the study and only one hospital investigated, so that our sample might not be considered representative of French inpatients' complaints. However, according to the data of the regional hospitalisation agency http://www.parhtage.sante.fr/re7/idf/site.nsf, the offer of care in our hospital may be considered representative of all teaching hospitals in Paris. Our study of complaints only captures an unknown fraction of patients that would complain of a medical error. As previously described [31], one reason is that "the complaint process is emotionally and financially costly, confusing, cumbersome and difficult to access." Since we analyzed only the written complaints arriving to the Department of Patient's Rights, a small number of complaints may not have been captured (e.g., those received by clinical departments and not transferred to the Department of Patient's Rights). We suppose that those non captured complaints were quantitatively and qualitatively minor, since in case of a major issue, patients would prefer that their complaint be managed by a hierarchic authority. Moreover, the staff of the Department of Patient's Rights is in charge of answering the patient's complaint and there may be no interest for this Department in underestimating the number of complaints received. Finally the design of this study did not allow to investigate, the factors influencing patients in actually deciding to file a complaint, neither could it state the causal relationship between the introduction of a new law and the number of complaints; nevertheless, we could clearly notice that the increase is contemporary to the implementation of the law.

Conclusion

In conclusion, this study reveals an increase in the number of complaints for medical issues in a university hospital in Paris over time, as well as an increase of the perception of a medical error after the diffusion of a law regarding patients' rights that may have contributed to this trend. Thus, we believe that the 2002 law regarding patient's rights may have contributed to an increase in complaints for medical issues due to a change in people's attitude towards the complaining process.

Competing Interests

The authors declare that they have no competing interests.

Authors' Contributions

CG and NG led the data analyses and the writing of the manuscript. VF, JJ and SET were involved in the study initiative and in data collection and preliminary analyses. JB participated actively in the management of databases and data analyses. GVT was the coordinator of the study and supervised all stages. All authors read and approved the final manuscript.

Authors' Information

GVT can be reached via the following email address: gwenaelle.vidal-trecan@ parisdescartes.fr

Acknowledgements

We are deeply indebted to Janine Vitiello and Pascale Finkelstein for their effort and support in this study. We are grateful to Hervé Dréau and Philippe Gnilka who provided activity data, Laure Albertini for her legal competence and Bruno Falissard for his statistical advices.

References

1. Agence Nationale pour l'Accréditation et l'Evaluation en Etablissement de Santé: Principes méthodologiques pour la gestion des risques en établissement de santé. Paris: Agence Nationale pour l'Accréditation et l'Evaluation en Etablissement de Santé; 2003.

2. Kohn LT, Corrigan JM, Donaldson MS: To err is human: building a safer health system. Washington, DC: National Academy Press; 1999.

3. World Health Assembly Resolution WHA55.18: Quality of care: patient safety. 55th World Health Assembly, Geneva, May 13–18, 2002 WHA55/202/REC/1[Volume 1: resolutions and decisions]. Geneva, World Health Organization; 2002.

4. Brennan TA, Leape LL: Adverse events, negligence in hospitalized patients: results from the Harvard Medical Practice Study. Perspect Healthc Risk Manage 1991, 11:2–8.

5. Michel P, Quenon JL, de Sarasqueta AM, Scemama O: Comparison of three methods for estimating rates of adverse events and rates of preventable adverse events in acute care hospitals. BMJ 2004, 328:199.

6. Michel P, Quenon JL, Djihoud A, Tricaud-Vialle S, de Sarasqueta AM: French national survey of inpatient adverse events prospectively assessed with ward staff. Qual Saf Health Care 2007, 16:369–377.

7. Davis P, Lay-Yee R, Briant R, Ali W, Scott A, Schug S: Adverse events in New Zealand public hospitals I: occurrence and impact. N Z Med J 2002, 115:U271.

8. Vincent C, Neale G, Woloshynowych M: Adverse events in British hospitals: preliminary retrospective record review. BMJ 2001, 322:517–519.

9. President's Advisory Commission on Consumer Protection and Quality in the Health Care Industry. Consumer Bill of Rights and Responsabilities. 17-7-1998 [http://www.hcqualitycommission.gov/] USA; Advisory Commission on Consumer Protection and Quality in the Health Care Industry 2009.

10. Starfield B: Is US health really the best in the world? JAMA 2000, 284:483–485.

11. President's Advisory Commission on Consumer Protection and Quality in the Health Care Industry. Quality First: Better Health Care for All Americans. 17-7-1998 [http://www.hcqualitycommission.gov/] USA; Advisory Commission on Consumer Protection and Quality in the Health Care Industry 2009.

12. Leape LL, Berwick DM: Five years after To Err Is Human: what have we learned? JAMA 2005, 293:2384–2390.

13. World Health Organization: WHO International Digest of Health Legislation. In Edited by World Health Organization. World Health Organization, Geneva; 2007.

14. Dew K, Roorda M: Institutional innovation and the handling of health complaints in New Zealand: an assessment. Health Policy 2001, 57:27–44.

15. Mitterand F, Beregovoy P, Dumas R, Vauzelle M, Joxe P, Teulade P, et al.: Loi modifiant le livre V du code de la santé publique et relative à la pharmacie et au. 1992.

16. Chirac J, Jospin L, Aubry M, Guigou E, Strauss-Kahn D, Le Pensec L, et al.: Loi relative au renforcement de la veille sanitaire et du contrôle de la securité sanitaire des produits destinés à l'homme. Loi n° 98-535. 1-7-1998. Code de la Santé Publique.

17. Chirac J, Jospin L, Fabius L, Guigou E, Lebranchu M-L, Vaillant D, et al.: Loi relative aux droits des malades et à la qualité du système de santé. Loi n° 2002-303. 4-3-2002. Code de la Santé Publique.

18. Friele RD, Sluijs EM: Patient expectations of fair complaint handling in hospitals: empirical data. BMC Health Serv Res 2006, 6:106.

19. Friele RD, Sluijs EM, Legemaate J: Complaints handling in hospitals: an empirical study of discrepancies between patients' expectations and their experiences. BMC Health Serv Res 2008, 8:199.

20. Parsons DW: Federal legislation efforts to improve patient safety. Eff Clin Pract 2000, 3:309–312.

21. Finlayson B, Dewar S: Reforming complaints systems: UK and New Zealand. Lancet 2001, 358:1290.

22. Teerawattananon Y, Tangcharoensathien V, Tantivess S, Mills A: Health sector regulation in Thailand: recent progress and the future agenda. Health Policy 2003, 63:323–338.

23. McCrindle J, Jones RK: Preliminary evaluation of the efficacy and implementation of the new NHS complaints procedure. Int J Health Care Qual Assur Inc Leadersh Health Serv 1998, 11:41–44.

24. Cunningham W, Crump R, Tomlin A: The characteristics of doctors receiving medical complaints: a cross-sectional survey of doctors in New Zealand. N Z Med J 2003, 116:U625.

25. Jarvis J, Frizelle F: Patients' complaints about doctors in surgical training. N Z Med J 2006, 119:U2026.

26. Tapper R, Malcolm L, Frizelle F: Surgeons' experiences of complaints to the Health and Disability Commissioner. N Z Med J 2004, 117:U975.

27. Andrews LB, Stocking C, Krizek T, Gottlieb L, Krizek C, Vargish T, et al.: An alternative strategy for studying adverse events in medical care. Lancet 1997, 349:309–313.

28. Wilson RM, Harrison BT, Gibberd RW, Hamilton JD: An analysis of the causes of adverse events from the Quality in Australian Health Care Study. Med J Aust 1999, 170:411–415.

29. Vincent C, Young M, Phillips A: Why do people sue doctors? A study of patients and relatives taking legal action. Lancet 1994, 343:1609–1613.

30. National Patient Safety Foundation: Public opinion of patient safety issues. In Edited by National Patient Safety Foundation. Chicago, IL, National Patient Safety Foundation; 1997.

31. Bismark MM, Brennan TA, Paterson RJ, Davis PB, Studdert DM: Relationship between complaints and quality of care in New Zealand: a descriptive analysis of complainants and non-complainants following adverse events. Qual Saf Health Care 2006, 15:17–22.

Three Methods to Monitor Utilization of Healthcare Services by the Poor

Abbas Bhuiya, S. M. A. Hanifi, Farhana Urni and
Shehrin Shaila Mahmood

ABSTRACT

Background

Achieving equity by way of improving the condition of the economically poor or otherwise disadvantaged is among the core goals of contemporary development paradigm. This places importance on monitoring outcome indicators among the poor. National surveys allow disaggregation of outcomes by socio-economic status at national level and do not have statistical adequacy to provide estimates for lower level administrative units. This limits the utility of these data for programme managers to know how well particular services are reaching the poor at the lowest level. Managers are thus left without a tool for monitoring results for the poor at lower levels. This paper demonstrates that with some extra efforts community and facility based data at the lower level can be used to monitor utilization of healthcare services by the poor.

Methods

Data used in this paper came from two sources- Chakaria Health and Demographic Surveillance System (HDSS) of ICDDR,B and from a special study conducted during 2006 among patients attending the public and private health facilities in Chakaria, Bangladesh. The outcome variables included use of skilled attendants for delivery and use of facilities. Rate-ratio, rate-difference, concentration index, benefit incidence ratio, sequential sampling, and Lot Quality Assurance Sampling were used to assess how pro-poor is the use of skilled attendants for delivery and healthcare facilities.

Findings

Poor are using skilled attendants for delivery far less than the better offs. Government health service facilities are used more than the private facilities by the poor.

Benefit incidence analysis and sequential sampling techniques could assess the situation realistically which can be used for monitoring utilization of services by poor. The visual display of the findings makes both these methods attractive. LQAS, on the other hand, requires small fixed sample and always enables decision making.

Conclusion

With some extra efforts monitoring of the utilization of healthcare services by the poor at the facilities can be done reliably. If monitored, the findings can guide the programme and facility managers to act in a timely fashion to improve the effectiveness of the programme in reaching the poor.

Background

Achievement of equity by way of improving the condition of the poor and disadvantaged in all aspects of life including health is one of the core goals of the contemporary development paradigm. It has been argued that unless performance indicators are examined by socioeconomic status of the population, improvement in average statistics may hide the presence of persistent or worsening inequities in a society [1]. This clearly indicates the need for monitoring the health and development indicators by socioeconomic status of the population. However, the challenge is to generate healthcare utilization data by socioeconomic status of the population with an acceptable level of statistical precision and reporting them regularly in an easily understandable fashion. National level healthcare utilization data often collected through cross sectional surveys, if analysed by the socioeconomic status of the population, can only portray the average level of disparities at the national level. Furthermore, this does not necessarily allow identification of inadequately performing regions, sub-regions, or lower level administrative or programme units with respect to reaching the poor. Thus, national level data serve a limited purpose for the facility/programme managers to assess the situation at the lowest level where most of the actions have to take place to improve the situation. Using routinely collected data from the facilities and communities through systems, such as, Health and Demographic Surveillance System (HDSS), utilization of healthcare services by the poor can be monitored at the local level. The 40 INDEPTH member sites in the developing world are uniquely placed to adopt the monitoring system to influence programmes and policies for enhancing utilization of health services by the poor [2]. Despite this potential, HDSS or similar other systems so far has put limited attention to monitoring utilization of health services by the poor at the local level. One of the reasons could be lack of attempts and demonstration of the methodological options available to do so. It is against this background that this paper is written.

Methods and Materials

The Study Area

The paper is based on data collected from Chakaria, a remote rural area in the south-east coast of Bangladesh. Chakaria is one of the 508 sub-districts in the country with a population of around 420,000. The area is a typical of rural Bangladesh with agriculture as the main occupation of its inhabitants. The infant mortality rate in the area during 2007 was 48 per 1,000 live births. Life expectancy was 69.7 years for females and 67.2 years for males. Total fertility rate was 3.5 children per woman. 95% of the deliveries during 2007 took place at home and only 19% of all the deliveries were assisted by skilled attendants [3]. The sub-district headquarters has one 31 bed primary care government hospital, three private clinics and an NGO hospital with inpatient and outpatient services. Primary health care services are provided through 13 primary healthcare centres run by the government. In addition, private services of nearly 40 physicians and 300 informal healthcare providers practicing modern medicine are available outside the institutional services [4].

Data Sources

Data collected from the households through quarterly visits as a part of Chakaria Health and Demographic Surveillance System (HDSS) during 2005–2007 and from government and private health facilities in Chakaria during March–June 2006 were used. Chakaria HDSS with its explicit focus on the poor and vulnerable, regularly collect data on ownership of household asset, occupation of main income earner and land owned by the household [3]. Chakaria is a member of INDEPTH network [2].

Categorization of Poor

Household socioeconomic status was assessed by asset score based on assets owned by any member of the household. The list of assets included television, radio, clock/watch, bedstead, phone, quilt, bi-cycle, wardrobe, and table/chair. In computing the asset index assets were assigned weights using the Principal Component Analysis (PCA) [5-7]. Despite some of its limitations, in many instances PCA has been recommended over the other alternatives for assigning weights in constructing asset index [8]. Households were categorized into quintiles based on the asset score. Proportion of households in the asset quintiles varied between 18–22 percent in the community. For the sequential sampling and LQAS households belonging to the lowest two asset quintiles were referred to as poor. In all other cases households from the lowest quintile were defined as poor.

Description of the Methods and their Operationalization

Two different approaches were used in monitoring utilization of skilled attendants for delivery and utilization of healthcare facilities for curative care by the poor. One was based on household level data and the other one on facility based data. Rates based methods used community based data and the Benefit Incidence, Sequential Sampling and LQAS used facility based data.

Rate-Ratio, Rate Difference, Concentration Curve and Concentration Index

Proportions of women who utilized skilled assistance during delivery from households in the five asset quintiles were computed for the years 2005 to 2007. Ratios and differences were calculated between the proportion in the lowest and the highest asset quintiles. Concentration index and concentration curve was constructed based on proportions from all the five asset quintiles. The concentration curve and related concentration index provides a means of assessing the degree of inequality in the distribution of a health variable. The value of concentration index can vary between -1 to +1 and a concentration index having a value of zero would indicate complete health equality among the various socioeconomic groups. On the other hand, a negative value would indicate a concentration of the health variable among the poorest group and a positive value would indicate the opposite [9-11]. The concentration index expresses the inequality in health across the full spectrum of socioeconomic status. In contrast, the rate-ratio and the rate-difference between the poorest and the richest quintile does not take into account the health status of the three middle quintiles [12].

Benefit Incidence Ratio

A one-week data, collected during March-May 2006, was plotted to analyse the Benefit Incidence Ratio. Asset scores were calculated for the patients by applying the same procedures as were done for the households. Information on ownership of assets similar to the one included in the HDSS was collected from the patients attending various facilities. The cut-off points for asset scores derived from the household level data were applied to the asset scores derived for the patients to categorize patients into quintiles. The proportions of patients in various quintiles were compared with 20% and any deviation from 20% would give an assessment of the extent to which the facilities have been serving the poor. This approach, commonly known as Benefit Incidence Ratio, has been in use for quite some time [13,14].

Sequential Sampling

Sequential sampling is commonly used for quality control in the industrial sector. In sequential test procedures the sample size needed to make a decision is not known in advance but rather determined by the sample results. In the sequential method, sample information is processed and evaluated as it becomes available, rather than at the end of the sampling process, as is done in fixed sample methods. The procedure continues to collect information only until enough evidence is available to make a decision confidently. The procedure was first developed by Wald (1947) [15]. The procedure uses a likelihood ratio to determine, after each observation is made, whether enough information is available to accept or reject the null hypothesis. Let us assume that L_1 represents the likelihood function of the sample result with k samples when the alternative hypothesis H_1 is true, and let L_0 represent the likelihood function when the null hypothesis H_0 is true. The ratio L_1/L_0 is the likelihood ratio. Details of likelihood function, null hypothesis and alternate hypothesis can be found elsewhere [16,17], When this ratio is large, the evidence points to H_1. When it is small, the evidence points to H0. Intermediate values are inconclusive. A sequential test can be performed by calculating L_1/L_0 after each new observation is available by applying the following (adopted from McWilliams [18]):

1. Stop with a reject H_0 decision if $L_1/L_0 > A$ (h_2+sk);
2. Stop with an accept H_0 decision if $L_1/L_0 < B$ (-h_1+sk); and
3. Continue to sample if $B \leq L_1/L_0 \leq A$.

Boundary values of A and B are chosen to satisfy Type I and Type II error specifications for the hypothesis test. Letting α and β represent probabilities of these errors respectively, A and B can be calculated according to

$$A = (1 - \beta)/\alpha, \quad B = \beta/(1 - \alpha).$$

The calculation of L_1/L_0 for each observation is tedious, but it can be shown mathematically that comparing L_1/L_0 to A and B for each observation is equivalent to comparing with h_2 + sk and -h_1 + sk respectively, where

$$r1 = \ln(p_1 / p_0), r2 = \ln[(1 - p_0)/(1 - p_1)]$$

$$a = \ln A = \ln\{(1 - \beta)/\alpha\}, b = -\ln B = \ln[(1 - \alpha)/\beta]$$

$$s = r_2/(r_1 + r_2), h_1 = b/(r_1 + r_2), h_2 = a/(r_1 + r_2).$$

In a plot of dk (cumulative number of non-conformities) versus k (observation) dk = -h_1+sk and dk = h_2+sk represent parallel lines, namely the "accept" and "reject" boundary lines. The test can be carried out by simply plotting dk versus k for each observation and continuing to sample until either the accept or the reject boundary is crossed and a decision is made. In practice, now-a-days, one can get the values calculated by using software and produce a table or a chart quite easily. Theoretically once the cumulative number of non-conformities falls in any one the two regions it can never change its direction and therefore it stays in that region irrespective of the number of additional non-conformities. For more details on sequential sampling one can consult Wald and McWilliams [15,18].

In our case the equivalent of non-conformities analogous to quality of industrial product was the number of patients from quintiles other than the lowest two quintiles. We performed the assessment at three levels of utilization by the non-poor: a) 20% as the lower limit and 40% as the upper limit (equivalent to 80% and 60% in terms of poor); b) 40% as the lower limit and 60% as the upper limit (equivalent to 60% and 40% in terms of poor); and c) 60% as the lower limit and 80% as the higher limit (equivalent to 40% and 20% in terms of poor). The calculation was done by using SISA software [19]. For instance, if we take the upper and lower boundaries of the poor patients based on 40% as lower threshold and 60% as upper threshold, it would mean if the proportion of poor attendees is more than 60% of the patients then the facility would be considered as serving the poor adequately. On the other hand, if the proportion of poor is less than

40% then the facility would be considered as inadequately serving the poor. If the proportion lies in between 40% to 60% then no decision about the adequacy/inadequacy of the facility in serving the poor could be made. This paper presents only the findings for the upper and lower thresholds for poor attendees at 60% and 40% respectively.

Data on ownership of assets included in the household survey were collected from the first and subsequent 99 patients attending the outdoor services in the sub-district public hospital and a private clinic. Data collection was stopped after interviewing 100 patients on a single day. Asset scores were calculated by applying the same procedure used in calculating asset scores for the households. The cut-off points for quintiles derived from the household survey were used in classifying the patients attending the facilities as poor. The cumulative number of non-poor patients attending the facility in a particular day was plotted against cumulative number of patients interviewed. The procedure stopped as soon as any of the boundary lines defining the rejection and acceptance regions, based on upper and lower thresholds for poor attendees at 60% and 40% respectively, was crossed.

Lot Quality Assurance Sampling (LQAS)

Lot quality assurance sampling (LQAS) originated in the manufacturing factory for quality control purposes to help the manufacturers in determining whether a batch or lot of goods can be accepted or rejected under pre-determined specifications [20]. In LQAS, a defective article is defined as one that fails to conform to the specifications of one or more quality characteristics. A common procedure in LQAS is to consider each submitted lot of product separately and to base the decision of acceptance or rejection of the lot on the evidence of one or more samples chosen at random from the lot [21].

Any systematic plan for single sampling requires that three numbers be specified. One is the number of articles 'N' in the lot from which the sample is to be drawn. The second is the number of articles 'n' in the random sample drawn from the lot. The third is the acceptance number 'd.' The acceptance number is the maximum allowable number of defective articles in the sample. More than 'd' defectives will cause the rejection of the lot. For instance, if we have a situation with N = 50, n = 5, and d = 0, it implies that "Take a random sample of size 5 from a lot of 50. If the sample contains more than 0 defectives, reject the lot; otherwise accept the lot." LQAS uses binomial probability to calculate the probability of accepting or rejecting a lot.

To apply the above in the context of monitoring utilization of health services by the poor, let us assume that the proportion of poor among the patients attending the facility is p. In a health facility with an infinitely large number of users, the probability P(a) of selecting a number a of poor in a sample size n is calculated as:

$$P(a) = \frac{n!}{a!(n-a)!} p^a q^{n-a}$$

where p = the proportion of poor attending the health facility,

q = (1-p),

n = the sample size,

a = the number of individuals in the sample who are poor,

n-a = the number of non-poor in the sample, usually denoted by d.

LQAS aids in choosing the sample size and the permissible value of n-a and interpreting the results. In order to use LQAS in the context of monitoring the utilization of a facility by the poor, the following five initial decisions must be made [22-24].

1. Firstly, the services to assess. This is selected by the health systems manager. In our case, let it be the attendance in the outdoor services.

2. Second, the facility to monitor (e.g., Upazila Health Complex (UHC), Union Health and Family Welfare Centre and the like).

3. Third, the target attendance to receive the services (e.g. any patient attending the facility, infants etc.).

4. Fourth, a triage system must be defined for classifying the level of usage by the poor as adequate, somewhat inadequate, and very inadequate. This needs to be decided by the programme managers, policy makers or other stakeholders related to the health service delivery.

5. Fifth, the levels of the provider and consumer risks (Provider risk is the probability of wrongly classifying a facility as very unsatisfactory which can put the reputation of the facility at risk; Consumer risk is the probability of wrongly classifying a very inadequately performing health facility as adequate which can put the poor in the area at health risk). In most cases it may be around 10–15%.

Using the information on the above five points, a series of operating characteristics (OC) curve (An OC curve depicts the probabilities of accepting a lot based on the proportion of non-conformance in the lot, the sample size, and the value of d, allowable non-conformances. An OC curve enables decision makers to examine the possible risks involved), or their corresponding probability tables can be constructed with the above binomial formula. From the OC curves, one can select the sample size (i.e. n) and the number of non-poor allowed (i.e. d) in the LQAS sample for a given level of provider and consumer risk before deciding that a health area has inadequate utilization by the poor.

Let us assume that a consensus has been reached among the various stake-holders of health service delivery in Bangladesh that facilities with 80% or more poor in their users can be considered as performing adequately. While facilities with 50% or less poor patients ought to be considered as very inadequately performing and be identified for attention. The ones in the mid-range 50% to 80% may be considered somewhat fine and for the time being they need no special attention. By using these information, probabilities of detecting "adequately performing" or "inadequately performing" health facilities can be calculated. Table 1 presents such probabilities along with provider and consumer risks for various combinations of sample sizes and maximum allowable non-poor patients in the sample.

Probabilities in Table 1 were calculated using the binomial formula. In each case, the upper and lower thresholds of the triage system were 80% and 50% respectively. The values in Table 1 (the row in bold) imply that in a sample of 28, if there are 9 or more non-poor, then the facility can be classified as inadequately performing in terms of serving the poor under the assumed triage of proportions (50%–80%) of poor. Details of LQAS method and its applicability in monitoring programme performance can be found elsewhere [24,25].

In our case, LQAS was applied in three scenarios with three levels of proportions of the poor in the facilities. In the first scenario, if the proportion of attendees in the facilities from the lowest two quintiles is less than 20%, then the facility is considered inadequate. If the proportion is more than 40%, then the facility is considered to be adequately performing. If the proportion is between 20%–40%, then no decision can be made. Under the above scenario, a facility can be considered as inadequately performing if in a sample of 50 attendees there are 35 or more are from quintiles other than the lowest two quintiles. The magnitude of misclassification in this case would be 11%.

In the second scenario, if the proportion of attendees from the lowest two quintiles is less than 40% then the facility is to be considered as inadequately performing in serving the poor. If the proportion is more than 60% then the facility is to be considered as adequately serving the poor. If the proportion is in between 40%–60% then no clear decision can be made. Under this scenario a facility can be considered as inadequately performing if in a sample of 50 patients, 25 or more are from quintiles other than the lowest two quintiles. The magnitude of misclassification in this case would be 16%.

Table 1. Example of application of the LQAS methodology

Sample size	No. in the sample non-poor	Probability of detecting health facilities with 80% poor as adequate	Probability of detecting health facilities with 50% poor as inadequate	Provider Risk	Consumer Risk	Total classification error
(n)	(d)	(a)	(b)	(1-a)	(1-b)	(1-a)+(1-b)
8	0	0.17	1	0.83	0	0.83
	1	0.50	0.96	0.50	0.04	0.54
	2	0.79	0.83	0.21	0.17	0.38*
	3	0.94	0.64	0.06	0.36	0.42
12	0	0.07	1.00	0.93	0.00	0.93
	1	0.28	1.00	0.73	0.00	0.73
	2	0.56	0.98	0.46	0.02	0.48
	3	0.80	0.93	0.21	0.07	0.28
	4	0.93	0.81	0.07	0.19	0.27*
	5	0.98	0.61	0.02	0.39	0.41
14	0	0.04	1	0.96	0	0.96
	1	0.20	1	0.80	0	0.80
	2	0.45	0.99	0.55	0.01	0.56
	3	0.70	0.97	0.30	0.03	0.33
	4	0.87	0.91	0.13	0.09	0.22*
	5	0.96	0.79	0.04	0.21	0.25
19	0	0.01	1	0.99	0	0.99
	1	0.08	1	0.92	0	0.92
	2	0.24	1	0.76	0	0.76
	3	0.46	1	0.54	0	0.55
	4	0.67	0.99	0.33	0.01	0.34
	5	0.84	0.97	0.17	0.03	0.20
	6	0.93	0.92	0.07	0.08	0.15*
	7	0.98	0.82	0.02	0.18	0.20
28	5	0.50	1	0.50	0	0.50
	6	0.68	1	0.32	0	0.32

Table 1. *(Continued)*

7	0.81	0.99	0.19	0.51	0.30
8	0.91	0.98	0.09	0.02	0.11
9	0.96	0.96	0.04	0.04	0.08*
10	0.99	0.90	0.01	0.10	0.11

* - Optimal decision rule for a sample size.
Source: Adopted from Valadez 1991, p.73.

The third scenario was with 60% as the lower and 80% as the higher thresholds. Under this scenario, a facility can be considered as inadequately serving the poor if in a sample of 50 there are 14 or more patients from other than the lowest two quintiles. The magnitude of misclassification in this case would be 11%. Although LQAS was applied in all these three scenarios, results based on 40%–60% thresholds are presented here. Decision regarding the pro-poor nature of the facilities could be made on a daily basis.

Findings

Rate-Ratio, Rate-Difference, Concentration Curve, and Concentration Index

Table 2 presents the use of skilled birth attendants for delivery in Chakaria during 2005–2007 by asset quintiles. It can be seen that the use of skilled assistance during delivery has increased overtime among women from households in all the quintiles except in the highest quintile. The absolute difference in the use of skilled attendant between highest and lowest quintile has reduced from 24 percent in 2005 to 11 in 2007. In relative term the ratio of percent of utilization among the women from the highest quintile and the women from the lowest quintile has reduced from six in 2005 to two in 2007. A similar picture of reducing inequities is seen when one compares the value of concentration index over time. Figure 1 visually depicts the reduction in inequities as reflected by the reduction in the areas between the concentration curve and the line of equality.

Figure 1 presents the concentration curves depicting the extent of inequalities in the use of skilled delivery assistance in Chakaria during 2005–2007. It can clearly be seen that the curves of inequality have been approaching the line of equality meaning a reduction in the level of inequalities over time.

Table 2. Percentage of women using skilled assistance during delivery, Chakaria 2005–2007

Asset quintile	2005	2006	2007
L(owest)	5.1	8.6	14.4
2	7.4	9.3	11.2
3	8.2	12.9	24.5
4	12.0	13.9	21.4
H(ighest)	29.6	28.6	25.4
Difference H-L	24.5	22.0	11.0
Ratio	5.8	3.3	1.8
Concentration Index	0.34	0.26	0.14

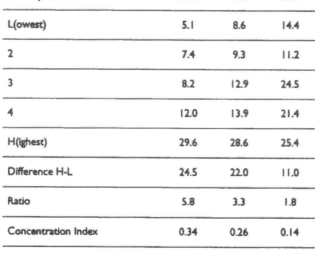

Figure 1. Concentration Curves of inequality in the use of skilled assistance during delivery, Chakaria 2005–2007. (Black line) – Line of equality (Blue line) – 2007 (Red line) – 2006 (Green line) – 2005

Benefit Incidence Ratio

Figure 2 and Figure 3 present distribution of patients attending a government facility and a private clinic respectively for outpatient services in Chakaria. The

line termed "community" parallel to the horizontal axis represents an equal distribution of services among the various quintiles in the community. Any deviation
from this line would indicate an unequal distribution of the services. Figure 2
shows that the patients in government facility were represented more by people
from the lowest quintile than they were in the community. The situation in the
private clinic was opposite of what was seen in the government facility (Figure 3).

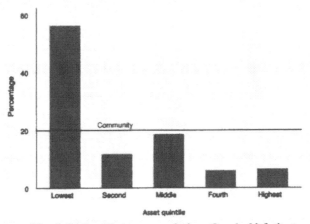

Figure 2. Application of Benefit Incidence Ratio to assess whether a Govt. health facility is used adequately by
the poor, Chakaria, March-May 2006.

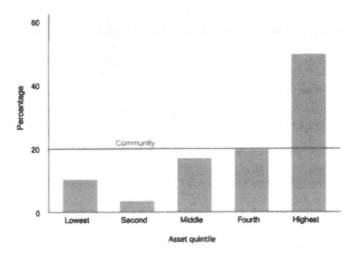

Figure 3. Application of Benefit Incidence Ratio to assess whether a private health facility is used adequately by
the poor, Chakaria, March-May 2006.

Sequential Sampling

Figure 4 and Figure 5 present results of the sequential sampling scheme for a government and a private facility in Chakaria respectively for the year 2006. It shows that the government facility was adequate in serving the poor with 40% and 60% as lower and upper thresholds of proportion of patients as poor and with 95% confidence level. These decisions for the government facility could be arrived at after interviewing the 42nd patient. While for the private facility it required interviewing only 10 patients to conclude that the facility was inadequate on that week as it had less than 40% of the patients from the lowest two quintiles.

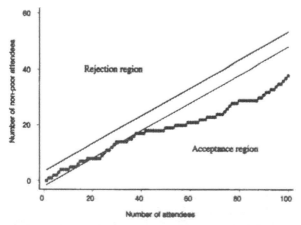

Figure 4. Application of Sequential Sampling Scheme to decide whether a Govt. health facility is used by the poor adequately, Chakaria, June 2006. Note: Results based on thresholds 40%–60%; alpha 5%; power 80%

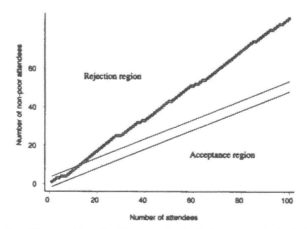

Figure 5. Application of Sequential Sampling Scheme to decide whether a private health facility is used by the poor adequately, Chakaria, June 2006. Note: Based on thresholds 40%–60%; alpha 5%; power 80%

Lot Quality Assurance Sampling

Table 3 presents number of non-poor patients found among the randomly chosen patients interviewed in the government and private facility. The application of LQAS in the Upazila Health Complex and the government facility resulted in terming the facility as serving the poor adequately or being pro-poor on all the six days of the interview week. The private clinic on the other hand failed in all the six days to be pro-poor. The exercise was based on 50 randomly chosen outdoor patients everyday for six days in a week. Of the 50 patients the maximum number of non-poor patients allowed was 25 with 40% and 60% lower and upper thresholds and 16% error of misclassification.

Table 3. Application of LQAS to assess whether health facilities are used by the poor adequately, Chakaria, June 2006

Period of evaluation Day	Threshold 40%–60%; Error = 16% Maximum number of non-poor (failure) permitted = 25			
	Upazila Health Complex		Private clinic	
	No. of non-Poor	Judgment	No. of non-Poor	Judgment
1	17	Pro-poor	42	Not pro-poor
2	13	Pro-poor	42	Not pro-poor
3	14	Pro-poor	45	Not pro-poor
4	13	Pro-poor	44	Not pro-poor
5	13	Pro-poor	43	Not pro-poor
6	21	Pro-poor	41	Not pro-poor

Note: based on information from 50 randomly chosen attendees

Discussion

The analysis of the utilization of skilled delivery attendants by women from various socioeconomic statuses as measured by asset scores was the most straightforward and familiar one to the demographers and epidemiologists. The extent of inequity could be summarized by rate-ratio and/or rate-difference. One of the caveats in this approach is that neither the rate-ratio nor the rate-difference makes use of the information for all the asset quintiles—they only make use of the utilization rates of the women from the lowest and highest quintiles. One of the ways to tackle this problem would be to use concentration index (CI), which is another way of measuring the degree of inequalities. CI makes use of the information from all the asset quintiles. The other limitation of these approaches is their inability to infer about the fixed health facilities. Although facility specific information can be collected during HDSS rounds or cross sectional surveys, quite often with many sources of health care services an analysis of this kind will demand a large sample size to make inference about a facility. Moreover frequency of such assessment has to be limited to the frequency of the rounds and the collection and analysis of the data may not be done by the facility managers.

The benefit incidence, sequential sampling and LQAS methods can be applied at the facility level to provide information about the pro-poor nature of the services and compared to the community based approaches these methods can be applied more frequently as desired by the facility managers or the researchers with far less effort. Moreover the sample size required for these three methods are relatively smaller compared to community based surveys [20,26]. Nevertheless, there are issues associated with each of these methods which are discussed below.

The benefit incidence analysis showed the over representation of the attendees from the lowest two quintiles at the government facilities and under representation at the private facilities. However, it did not resort to formal statistical hypothesis testing in terms of identifying how big a deviation from 20% should be of concern. One can, of course, compare the proportions in the facility with 20% by using statistical tests. Such tests, however, would require the denominators from which the community proportion and the facility proportion were derived. In addition, a computation of the test statistics and associated probability to make an inference about the difference between the proportions of patients from the lowest quintiles would also be warranted. In case of LQAS and sequential sampling plans, the issue regarding how big a difference would be of significance is embedded in the procedure. In effect, the procedures operationalized those formal statistical testing in terms of number of non-conforming attendees, which in these cases, were from the quintiles other than the lowest two quintiles, with predetermined levels of error and power. The sequential plan has the advantage of plotting the cumulative number of non-poor against the number of attendees assessed for their socioeconomic status (SES) as they come, and provide a powerful visual tool for the facility managers. To have an equivalent visual representation in LQAS may not be that straightforward. Benefit incidence analysis, however, has the advantage of visual presentation without the formal statistical inference procedures built in. The caveat in the sequential sampling plan is that in some instances it may lead to a large number of interviews before a facility can be validly classified as pro-poor. This issue has been addressed in LQAS. LQAS combines the sequential test procedures with a fixed sample scheme in the sense that it allows decision-making by testing a fixed number of cases with a predetermined level of error.

In a situation leading to a non-stop examination of cases under sequential sampling, one can also resort to double sampling, meaning that if sequential sampling does not enable a decision making after examining a sample of cases then one can take another sample. However, more than two samples do not provide any additional advantage. LQAS has taken care of this issue of not being able to make a decision, for it combines sequential sampling and fixed sample methods. In case of LQAS, as we have seen, the number of attendees to be included in the sample is predetermined given the level of errors and thresholds, and thus it avoids the situation of no decision making. Methodologically speaking, the sequential plan and LQAS are almost similar with the above weaknesses and strengths. Either of them would serve the purpose of drawing inferences about the pro-poor nature of the services in terms of utilization by the poor. Facility management staff members can easily be trained to adopt any of these two methods.

Another challenge is the identification of the poor. We used asset quintiles for it allows the classification of attendees in terms of any interval such as deciles or quintiles, and in particular, allows the identification of the bottom twenty percent of the population. The challenge is to train facility managers to identify attendees from the lowest quintiles. This requires values for weights of assets and cut off points of asset scores based on the distribution of households in the community. Thus a community survey or an approximation from other surveys is required. Once the cut-off points are known, then the facility managers have to be trained in how to use the weights in calculating asset scores for the attendees, and how to use the cut-off points to identify attendees from the lowest quintiles. Easier alternatives exist that are simpler than using asset scores. These include using the number of assets owned, or other indicators such as land, occupation of main income earner, level of education and the like. The challenge in using these is to get deciles and quintiles. Use of indicators other than asset scores would obviously make the adoption of the monitoring system very attractive.

Another practical issue one has to deal with in adopting these methods is to decide how frequently the assessment should be made or, in other words, how frequently the data at the facility and the community level should be collected. The answer to the frequency of data collection at the community level is somewhat dependent on the chances of changes in the SES of the community. In many instances, the changes in SES are slow. The frequency of assessment at the facility level is dependent on the facility managers to some extent and also on the nature of services to be assessed. Again, it will largely depend on the nature of changes in the services or in the system. If the system is stable in terms of design, then perhaps, it is not useful to have very frequent assessments. If there is a special service for a short period and it is very important to make the service responsive to every section of the society, then perhaps it would be useful to increase the frequency of monitoring. The other issue to consider in deciding frequency of monitoring is the presence of a pattern during certain days, weeks, or months of the year when the facility is used by certain segments of the society more than usual. If such is the case, then this information should be used in deciding the timing and frequency of assessments. It may be mentioned that in the two upazilas where we worked, we examined the variation in use of the facilities by the SES of the attendees, and in most cases, no significant statistical variation was observed. This means that any day of the month would represent the pattern of the whole month satisfactorily.

Conclusion

Benefit incidence analysis can be a starting point for monitoring the utilization of health services by the poor. Sequential sampling scheme allows a more formal inference about the performance in terms of utilization of services by the poor and its visual display of the findings on a continual basis makes the procedure quite attractive. LQAS may be preferable to sequential sampling for its ability to make a decision within a fixed sample size. Finally, HDSS sites or other data collection systems with data on socioeconomic status can examine health care utilization by the poor or any other marginalized group to assess how well health care services reach the poor and disadvantaged. This can guide the facility and programme managers to take appropriate actions in a timely fashion to ensure programmes reach the targeted population.

Abbreviations

HDSS: Health and Demographic Surveillance System; UHC: Upazila Health Complex; LQAS: Lot Quality Assurance Sampling; OC: Operating Characteristics; CI: Concentration Index; SES: Socioeconomic Status.

Competing Interests

The authors declare that they have no competing interests.

Authors' Contributions

AB conceived the study, designed and implemented it. He also supervised data collection, analysis and drafted the manuscript. SMAH contributed in study design, supervised data collection and conducted the data analysis. He also contributed in drafting the manuscript. FU contributed in data analysis. SSM contributed in literature review for the manuscript and revising the manuscript. All authors read and approved the final manuscript.

Acknowledgements

Data presented in this paper were collected through the Chakaria Health and Demographic Surveillance System, one of the activities of the Chakaria Community Health Project, maintained by ICDDR,B. This research study was funded by ICDDR,B, Department for International Development (DFID), UK, through the 'Future Health Systems: Innovations for equity—a research program consortium, grant number GR-00445 and by the Government of Bangladesh through IHP-HNPRP, grant number GR-00410. ICDDR,B acknowledges with gratitude the commitment of DFID and the Government of Bangladesh to the Centre's research efforts. ICDDR,B also gratefully acknowledges the following donors which provide unrestricted support to the Centre's research efforts: Australian Agency for International Development (AusAID), Government of the People's Republic of Bangladesh, Canadian International Development Agency (CIDA), Embassy of the Kingdom of the Netherlands (EKN), Swedish International Development Cooperation Agency (Sida), Swiss Agency for Development and Cooperation (SDC), and Department for International Development, UK (DFID).

The authors would further like to acknowledge the support of the community people and the authorities of the health service facilities monitored under the study.

References

1. Gwatkin DR: How much would poor people gain from faster progress towards the Millennium Development Goals for health? Lancet 2005, 365(9461):813–817.

2. INDEPTH Network [http://www.indepth-network.org/]

3. Bhuiya A, Hanifi SMA, Urni F, Iqbal M: Chakaria Health and Demographic Surveillance System: Focusing on the poor and vulnerable. Demographic events and safe motherhood practices-2007. In Scientific report no 105. Dhaka: ICDDR, B; 2008:47.

4. Bhuiya A, ed: Health for the rural masses: Insights from Chakaria. Dhaka: International Centre for Diarrhoeal Disease Research, Bangladesh; 2009.

5. Filmer D, Pritchett L: Estimating wealth effect without expenditure data- or tears: An application to educational enrolments in states of India. Demography 2001, 38:115–132.

6. Gwatkin DR, Rustein S, Johnson K, Suliman EA, Wagstaff A: Socio-economic differences in health, nutrition, and population. In Technical report, World Bank. Volume 1. Armenia- Kyrgyz Republic; 2003:400.

7. Hotelling H: Analysis of a complex of statistical variables into principal components. J Educ Psychol 1933, 24:417–441.

8. Howe LD, Hargreaves JR, Huttly SR: Issues in the construction of wealth indices for the measurement of socio-economic position in low-income countries. Emerg Themes Epidemiol 2008, 5:3.

9. Kakwani NC, Wagstaff A, Doorlsaer EV: Socioeconomic inequalities in health: Measurement, computation and statistical inference. J Econom 1997, 77(1):87–104.

10. Pan American Health Organization (PAHO): Measuring Inequalities in Health: Gini Coefficient and Concentration Index. Epidemiological Bulletin 2001, 22(1):3–4.

11. Wagstaff A, Paci P, Van Doorslaer E: On the measurement of inequalities in health. Soc Sci Med 1991, 33:545–557.

12. Anand S, Diderichsen F, Evans T, Shkolnikov VM, Wirth M: Measuring disparities in health: Methods and indicators. In Challenging inequities in health: from ethics to action. Edited by: Evans T, Whitehead M, Diderichsen F, Bhuiya A, Wirth M. New York: Oxford University Press; 2001.

13. Gasparini L, Panadeiros M: Argentina: Assessment of changes in the distribution of benefits from health and nutrition policies. In Reaching the Poor with Health Services. Edited by: Gwatkin D, Wagstaff A, Yazbeck A. The World Bank; 2005.

14. Bhuiya A, Hanifi SMA, Chowdhury M, Jahangir M, Gwatkin DR: Rapid Methods for monitoring the utilization of healthcare facilities by the poor: Findings from a pilot project in rural Bangladesh. Future Health Systems Working Paper No 2, Bangladesh Series 2007.

15. Wald A: Sequential Analysis. New York: John Wiley Sons, Inc; 1947.

16. Young GA, Smith RL: Essentials of Statistical Inference. Cambridge University Press; 2005.

17. Harris JW, Stocker H: Handbook of mathematics and computational science. New York: Springer-Verlag; 1998.

18. McWilliams TP: How to use sequential statistical Methods. In The ASQC basic references in quality control. Volume 13. Milwaukee, WI.: ASQC Quality Press; 1988.

19. SISA-Binomial [http://www.quantitativeskills.com/sisa/distributions/binomial.htm]

20. Hoshaw-Woodard S: Description and comparison of the Methods of cluster sampling and lot quality assurance sampling to assess immunization coverage. Geneva: Department of Vaccines and Biologicals, WHO; 2001:15.

21. Grant EL, Leavenworth RS: Statistical quality control. 5th edition. Singapore: McGraw-Hill; 1998.

22. Rosero-Bixby L, Grimaldo C, Raabe C: Monitoring a primary health care programme with lot quality assurance sampling. Health Policy Plan 1990, 5:30–39.

23. Valadez J, Bamberger M: Monitoring and Evaluating Social Programs in Developing Countries: A Handbook for Policymakers, Managers, and Researchers. Washington, DC: World Bank (EDI development studies); 1994.

24. Valadez JJ: Assessing child survival programs in developing countries: Testing lot quality assurance sampling. Boston: Harvard University Press; 1991.

25. Bhuiya A, Hanifi SMA, Roy N, Streatfield PK: Performance of the lot quality assurance sampling method compared to surveillance for identifying inadequately-performing areas in Matlab, Bangladesh. J Health Popul Nutr 2007, 25(1):37–46.

26. Myatt M, Bennett DE: A novel sequential sampling technique for the surveillance of transmitted HIV drug resistance by cross-sectional survey for use in low resource settings. Antivir Ther 2008, 13(Supplement 2):37–48.

Association between Community Health Center and Rural Health Clinic Presence and County-Level Hospitalization Rates for Ambulatory Care Sensitive Conditions: An Analysis Across Eight US States

Janice C. Probst, James N. Laditka and Sarah B. Laditka

ABSTRACT

Background

Federally qualified community health centers (CHCs) and rural health clinics (RHCs) are intended to provide access to care for vulnerable populations.

While some research has explored the effects of CHCs on population health, little information exists regarding RHC effects. We sought to clarify the contribution that CHCs and RHCs may make to the accessibility of primary health care, as measured by county-level rates of hospitalization for ambulatory care sensitive (ACS) conditions.

Methods

We conducted an ecologic analysis of the relationship between facility presence and county-level hospitalization rates, using 2002 discharge data from eight states within the US (579 counties). Counties were categorized by facility availability: CHC(s) only, RHC(s) only, both (CHC and RHC), and neither. US Agency for Healthcare Research and Quality definitions were used to identify ACS diagnoses. Discharge rates were based on the individual's county of residence and were obtained by dividing ACS hospitalizations by the relevant county population. We calculated ACS rates separately for children, working age adults, and older individuals, and for uninsured children and working age adults. To ensure stable rates, we excluded counties having fewer than 1,000 residents in the child or working age adult categories, or 500 residents among those 65 and older. Multivariate Poisson analysis was used to calculate adjusted rate ratios.

Results

Among working age adults, rate ratio (RR) comparing ACS hospitalization rates for CHC-only counties to those of counties with neither facility was 0.86 (95% Confidence Interval, CI, 0.78–0.95). Among older adults, the rate ratio for CHC-only counties compared to counties with neither facility was 0.84 (CI 0.81–0.87); for counties with both CHC and RHC present, the RR was 0.88 (CI 0.84–0.92). No CHC/RHC effects were found for children. No effects were found on estimated hospitalization rates among uninsured populations.

Conclusion

Our results suggest that CHCs and RHCs may play a useful role in providing access to primary health care. Their presence in a county may help to limit the county's rate of hospitalization for ACS diagnoses, particularly among older people.

Background

Rural Safety Net Providers

Access to primary health care in the US is affected by an individual's financial ability to pay for care, principally measured by insurance, and by the availability of a practitioner to provide services. Access in many rural counties is challenged at both the individual and the facility level: rural areas have proportionately more poor and uninsured persons than urban areas, and are served by fewer health care providers. [1,2] A number of urban counties are similarly at risk. [3] In both rural and urban settings, safety net facilities can have marked effects on population health. Two principal types of federally designated safety net facilities serve these areas: federally qualified community health centers (CHCs) and rural health clinics (RHCs). CHCs and RHCs are located in counties with demonstrated high need for care among at risk populations, and those that have been designated as rural, respectively.

Community health centers, administered by the Bureau of Primary Care, Health Resources and Services Administration (HRSA), have been the principal Federal vehicle for providing health care access to poor and uninsured persons. CHCs, which must be located in a medically underserved area, receive Federal grant funding that allows them to care for patients of limited financial means and to provide expanded services, such as transportation assistance, for vulnerable groups. Based on HRSA data, CHCs provided care for more than 15 million individuals in 2006, of whom nearly two thirds were of minority race/ethnicity. [4] Most CHC clients were at or below poverty (71%) and a substantial minority were uninsured (40%). [5] CHCs must accept all patients regardless of ability to pay, with a sliding-fee scale for the poor and uninsured. However, CHCs are expected to be "financially viable and cost-competitive;" thus, they are not required to provide free care to all patients. [6]

The Rural Health Clinic (RHC) program is directed toward the retention of physicians and other providers in rural areas. Established in 1977, it allows participating medical practices to receive higher reimbursement from Medicare and Medicaid, major payers for rural populations. [7] RHCs must be located in non-metropolitan Health Professional Shortage Areas (HPSAs), either a geographic shortage area (where the entire county lacks providers), or a population group shortage area (where specific types of individuals are underserved). Because the definition used for "rural" may either follow Federal guidelines or be set by a state governor, rural HPSAs can exist in counties that are classified as metropolitan or urban by the US Census. RHCs are not required to provide a full spectrum of primary care services; nor are they required to see all individuals seeking care regardless of need. As of 2005, 16 percent (590/3600) of RHCs stated that they would take all patients regardless of insurance status. [8] Although not required to accept uninsured individuals, RHCs actually derive a greater proportion of practice revenue from uninsured patients than do CHCs (15% versus 7%). [9] Advocacy groups, such as the National Rural Health Association, consider the

RHC program a safety net function because of its role in rural physician recruitment and retention.

Assessments of CHC and RHC Effects on Population Health

We sought to clarify the contribution that CHCs and RHCs may make to the accessibility of primary health care, as measured by rates of hospitalization for ambulatory care sensitive (ACS) conditions. ACS conditions are those for which, in the consensus of medical experts, primary care of acceptable quality can reduce the frequency of hospitalization. [10-16]

Ambulatory Care Sensitive Hospitalizations as a Measure of Access

ACS hospitalization as an indicator of primary care access assumes that quality outpatient care, by linking the patient to effective assessment, education, pharmacological management, and other treatment, reduces the likelihood that patients with specific diagnoses will need hospitalization. [10,15,16] While all relevant hospitalizations cannot be prevented, at the population level ACS hospitalizations have been found to be lower where other measures of access to care, such as provider availability, are higher. [17,18] Hospitalization rates for ACS conditions are higher in rural areas [19-22] and among nonwhites and individuals with low incomes. [23] ACS hospitalization rates are used by the Agency for Healthcare Research and Quality to measure access among minority populations, and in general assessments of safety net performance. [24]

Evidence for CHC and RHC Effects on Population Health

Prior research has found that individuals insured by Medicaid who received most of their care at a CHC, compared with another single facility, were less likely to be hospitalized or to visit an emergency room for ACS conditions. [25,26] The presence of a CHC in a medical market area has been associated with lower ACS admission rates. [27] At the county level, the presence of a CHC has been shown to reduce ACS hospitalization rates among children. [28] Given this previous research, we anticipate that population-based hospitalization rates in counties served by CHCs will be lower than in counties lacking these facilities.

RHCs may also have effects on population health, although research in this area is sparse. Evidence suggests that an RHC can be financially beneficial to a sponsoring hospital. [29]. A person-level analysis, limited to Nebraska, found that hospitalizations in whole-county HPSAs containing an RHC were less likely to involve an ACS condition than those in counties without an RHC. [30] This study, which did not examine population-level risks for ACS hospitalization, is the only previous research examining the association between the presence of an RHC and population health.

Our analysis expands on previous work by examining effects of CHCs, RHCs, and both facilities in combination, on population level access to care, as measured by county-level hospitalization rates for ACS conditions. Our primary analysis examines these rates in the general population, stratified by age. Because CHCs have a specific mission to help medically indigent populations, including uninsured persons, we also examine the association between the presence of CHCs and/or RHCs and ACS admission rates among the uninsured.

Methods

Sample

We used a cross-sectional, ecologic design to explore relationships between county-level rates of ACS diagnoses and county level covariates, including the presence of a CHC, RHC, or both. ACS hospitalization rates are calculated based on hospitalizations of persons who reside in the county, regardless of where the hospitalization takes place. The unit of analysis is an age- and county-specific rate, calculated based on discharges of county residents. Counties were used, rather than smaller geographic units, because most of the data elements needed for multivariate analysis are available only at this level. The study was approved by the Institutional Review Board of the University of South Carolina.

Data were drawn from the 2002 State Inpatient Databases (SIDs). The SIDs, compiled at the state level and supported for research use by the Agency for Healthcare Research and Quality, contain discharge records for all hospitalizations in participating states (100 percent). Only fifteen states included information about patients' counties of residence in 2002. Budgetary constraints coupled with the per-state cost for SID files limited the analysis to eight states: Colorado, Florida, Kentucky, Michigan, New York, North Carolina, South Carolina, and Washington. These states were chosen to provide at least one state in each of the four major Census Divisions of the US, and to offer a large number of counties with CHCs, RHCs, or both facilities. The presence of a CHC or RHC in each county was determined using Area Resource File data for the year 2002. Four mutually exclusive categories were created: CHC but no RHC, RHC but no CHC, both (CHC and RHC), and neither facility. Across all counties, 59 (10.2%) had a CHC and not an RHC; 139 (24.0%) had an RHC but not a CHC; 27 had both facilities (4.7%), and 354 (61.1%) counties had neither facility. Counties were distributed by state as follows: Colorado, 63 counties (10.9%); Florida, 66 counties (11.4%); Kentucky, 120 counties (20.7%); Michigan, 83 counties (14.3%), New York, 62 counties (10.7%); North Carolina, 100 counties, 17.3%; South Carolina, 46 counties (7.9%), and Washington, 39 counties (6.7%).

Measurement of ACS Conditions

We used definitions for ACS diagnoses from the Agency for Healthcare Research and Quality. [31] The ACS conditions for adults are specific diagnoses for asthma, angina (without procedure), congestive heart failure (CHF), bacterial pneumonia, chronic obstructive pulmonary disease (COPD), dehydration, diabetes long-term complications, diabetes short-term complications, hypertension, lower-extremity amputation for individuals with diabetes, perforated appendix, uncontrolled diabetes, and urinary tract infection. For children, the ACS conditions include asthma, bacterial pneumonia, dehydration, perforated appendix, gastroenteritis, and urinary tract infection. The precise definitions used in this research account for a variety of exclusions detailed in technical specifications that are readily available from the AHRQ [31]. For children, for example, hospitalizations for asthma are excluded if there is evidence of cystic fibrosis or anomalies of the respiratory system. A hospitalization for any of these diagnosis is considered to be a hospitalization for an ACS condition. We did not attempt to study hospitalizations for individual diagnoses because of the instability of rates in counties with very small populations, and also because most research in this area uses the combined indicator.

Analytic Approach

We examined adjusted rates of ACS hospitalization in counties with a CHC, an RHC, or both, and compared these to the analogous rates in counties with neither facility. We calculated ACS rates separately for children (0–17), working age adults (18–64), and older individuals (65 and over). To ensure stable rate estimation, we established population-based criteria for county inclusion before conducting data analysis. County-level population estimates for 2002 were drawn from the 2005 Area Resource File (n = 579 counties). We included a county in the rate analysis for children and for working age adults only if it had at least 1,000 persons ages 0 – 17 (children) or ages 18 – 64 (working age adults). For age 65 and over, we included a county only if it had at least 500 persons in that age group; the threshold of inclusion was lower for older persons because of their higher ACS admission rates. These criteria excluded 21/579 counties from the analysis for children (3.6%), 5/579 counties from the analysis for working age adults (0.9%), and 12/579 counties from the analysis for older adults (2.1%). We considered an alternative approach, retaining all counties in the analysis and adjusting standard errors to account for heteroskedasticity. We judged that this approach might not adequately account for unrepresentative high or low ACSH rates that could appear among such small populations. Such unrepresentative rates could introduce bias into the estimations, because they could be attributable to even small random variations in the number of individuals hospitalized for ACSCs in these small populations, rather than to differences in access to primary health care. A comparison of mean county population and mean number of ACS discharges for included and excluded counties is provided in Table 1.

Table 1. Mean Number of Persons in Each Age Range for Included and Excluded Counties, and Mean Number of ACSC Hospitalizations in Each Age Range in these Counties

	Ages 0–17		
	Number of Counties	Mean Population	Mean Number of ACSC Discharges
Included Counties	559	31,535	152.4
Excluded Counties	20	614	1.8
County Total	579		
	Ages 18–64		
Included Counties	574	78,778	687.1
Excluded Counties	5	661	4.6
County Total	579		
	Ages 65+		
Included Counties	Number of Counties	Mean Population	Mean Number of ACSC Discharges
Excluded Counties	567	16,912	1,111
County Total	12	276	14.7
	579		

As noted in the introduction, CHCs and RHCs are located only in specific county types and are not randomly distributed across the US. As illustrated in Table 2, counties with CHCs and/or RHCs differ from counties in those same states with neither facility in several characteristics, including HMO penetration and proportion of the population that is uninsured. To adjust for differences between studied counties and counties with neither facility, adjusted analyses controlled for the county characteristics listed in Table 2.

Table 2. Characteristics of counties, by CHCs/RHCs in the county, studied states, 2002.

SID Sample, n = 579	Counties in studied states only, white				All U.S. Counties
	CHC Only	RHC Only	Both CHC and RHC	Neither facility	
Number of Counties:	59	139	27	354	3,168
Resources in county:					
MD/DO per 10,000 population	12.9	10.3	14.3	12.3	12.1
Beds per 10,000 population	3.6	3.2	3.9	3.2	3.9
Number of hospitals with emergency department	1.7	1.0	1.4	1.6	1.3
HMO penetration rate	25.3 ***	9.4	10.6	14.2	11.4
ED visits per 1,000 population	337	372	382	330	351
Non-metropolitan county (%)	23.7 ***	79.9	66.7 ***	58.8	65.3
Characteristics of county populations					
Percent of population that is:					
African American	20.3	17.5	16.2	16.4	9.5
Hispanic white	6.4	6.6	7.7	6.4	5.3
Asian	1.7	1.6	3.5	2.1	1.0
American Indian/Native American	2.1	5.1	2.1	2.3	1.9
Population change, 1990 – 2000 (%)	10.4	12.5	8.5	13.0	8.1
Percent of population with less than a high school education	24.9	24.3	24.0	25.0	22.6
Population per square mile	167	141	183	219	23
Percent of population that is unemployed	7.3	6.8	6.3	7.0	7.1
Percent uninsured, aged 18–64	19.0	20.8	22.1 **	18.9	19.6
Percent uninsured, age 17 or less	12.0	12.9	13.3 *	11.4	12.4
Median household income	35,595	35,179	36,835	35,844	35,363
Death rate per 10,000 due to:					
Cardiovascular disease	18.3	15.6	17.1	17.0	20.7
Chronic obstructive pulmonary disease	5.1	4.6	5.1	4.7	5.1
Diabetes	2.6	2.2	2.2	2.2	2.8
Liver disease	1.0	0.9	0.9	0.9	0.9

Data Source: Authors' analysis using year 2002 State Inpatient Databases (8 states), and the 2002 Area Resource File (ARF). Note that ARF death rates are based on 3-year average. Statistical tests compare the indicated category to counties that have neither facility type; statistical tests are t-tests, except for the chi-square test for non-metropolitan counties (conducted as a single chi-square for all CHC/RHC combinations). N=3168 is the number of US counties in states. Excludes ARF counties from U.S. Territories (Guam, Puerto Rico, etc.).
***p < .001; **p < .01; *p < .05

The models for this study are based on Andersen's (1995) conceptualization of use of health services as resulting from the multiple influences of the external community and health services environment, population characteristics, health behavior, and outcomes. [32] Variables representing health system characteristics and use included physician supply, bed supply, number of hospitals with an emergency department, emergency department visit rates, and managed care penetration rates. Physician supply is generally inversely related to ACS hospitalization rates [17,20,33], but a positive relationship [34] and no relationship [35,36] have also been found. Managed care penetration has been found to be inversely related to ACS hospitalization rates. [37,38] County characteristics measured included racial/ethnic composition of the population (proportions that are non-Hispanic black, Hispanic, Asian American, and American Indian/Native American), population change 1990 – 2000; the percent of the population with less than a high school education, the unemployment rate, population per square mile, and whether the county was classified as metropolitan (urban) or non-metropolitan (rural). [39] The racial/ethnic composition of the population is included to adjust for differing patterns of health and health care use among minorities. [17,40-42] Population change, education levels, and unemployment are used as measures of the financial and economic status of the county as a whole. Population density is used, in addition to rural status, to adjust for differences within rural counties. Including a rural/urban variable in the model does not introduce unacceptable colinearity with the covariate representing RHCs, because a notable proportion of counties with RHCs are classified as metropolitan (Table 2). Resource characteristics included median household income and the percent of the population estimated to lack health insurance. Estimates of the uninsured population in each county were obtained from the U.S. Census. [43] Consistent with previous research, we included four covariates to control for county health burdens: unadjusted death rates from cardiovascular disease, chronic obstructive pulmonary disease, diabetes, and liver disease. [32] Table 2 provides a full description of these parameters across county types. With the exception of county-level estimates of the uninsured population, all variables are drawn from the Area Resource File.

Multivariate Poisson analysis was used to calculate adjusted rate ratios comparing counties with one or more CHCs, one or more RHCs, or at least one CHC plus at least one RHC, to counties having none of these facility types, while holding other county characteristics equal. The rate ratio is the ratio of the mean value of ACS hospital admission rates across counties of a given type, separately estimated for each age group, where the mean rate for a county type of interest (such as counties with both a CHC and an RHC) is the numerator. The denominator is the corresponding rate for counties having neither a CHC nor an RHC, the reference category. The rate ratio is obtained by exponentiating the estimate of interest from the Poisson analysis. Rate ratios less than 1.00 suggest that the hospitalization rate in the county type of interest was lower than the rate in the reference category.

For calculating rates among uninsured adults, we used Census estimates of the number of uninsured adults in each county as the denominator. We made the assumption that nearly all such persons are younger than 65, as most older people are covered by Medicare. For the separate analysis of children, the denominator was the Census estimate of uninsured children. The numerator specific to each age group in each county was the number of ACS admissions for which the payment source was identified as "self pay" in the discharge record. This value may not precisely equal the uninsured population, as some self-pay admissions may later have been converted to an insurer; however, it is reasonable to assume that the number of cases in which this occurred is relatively small. Measurement errors, if present, might have the greatest effect on ACS admission rates among children, which are generally quite low and thus could be affected by small changes.

Results

ACS Hospitalization Rates Across County Populations

Unadjusted ACS hospitalization rates were lowest among children and markedly higher in the 65 or older population. Unadjusted ACS rates among children did not differ by CHC/RHC availability. ACS hospitalization rates in the working age and age 65 or above populations were significantly lower in counties with a CHC than in counties with neither facility; rates in counties with an RHC only, or both facilities, did not differ from those in counties with neither facility.

In adjusted analysis, the presence of a CHC or RHC in the county was associated with ACS hospitalization rates for children only for the comparison of counties with both a CHC and RHC with those having neither facility. The rate ratio comparing these counties, 1.30 (95% Confidence Interval, CI 1.10–1.55), suggests that ACS hospitalizations are more common in counties with both facility. Among working age adults, the ACS hospitalization rate in counties having a CHC was 0.86 of the rate in counties with neither facility type (95% CI 0.78–0.95). ACS hospitalization rates in counties with an RHC only, or with both facility types, did not differ from those in the comparison group. Among older adults, counties with either safety net facility had lower ACS hospitalization rates than counties with none. The rate in counties with one or more CHCs, but no RHC, was 16% lower than those with neither facility type (rate ratio, RR, 0.84, CI 0.81–0.87). The rate in counties with one or more RHCs, but no CHC, was 4% lower than that in counties with neither facility type (RR 0.96, CI 0.94–0.99). The rate in counties with at least one CHC and at least one RHC was 12% lower than in those with neither facility type (RR 0.88, CI 0.84–0.92).

We examined the residuals from these analyses to identify whether some states or county clusters might systematically exhibit an association between CHCs/ RHCs in the opposite direction from these estimated results. Although ACS hospitalization rates were generally higher in Kentucky than in the other states in the analysis, with greater variation in these rates in Kentucky as well, there was no indication that the rates might systematically depart in direction from these adjusted averages for particular states or clusters of counties.

Multiple county characteristics in addition to CHC/RHC presence were associated with ACS hospitalization rates, particularly among older adults. Factors with similar effects across all age groups included HMO penetration (lower rates), positive population change (lower rates), and cardiovascular disease death rates (greater rates). Residence in a non-metropolitan county was associated with notably lower ACS hospitalization rates among children, and modestly lower rates among older adults, with facility availability held constant.

ACS Hospitalization Rates Among Estimated Uninsured County Populations

There was no evidence that the presence of a CHC or RHC was associated with lower ACS hospitalization rates for uninsured children (Tables 3 and 4). In unadjusted results for uninsured working age adults, counties with at least one CHC, but no RHC, had an average ACS hospitalization rate per 1,000 uninsured persons of 8.44, compared with an average rate of 10.40 for counties with neither safety net facility (p = 0.0029). When demographic and health resource characteristics of the counties were controlled in multivariable analysis, there were no differences in the rates of ACS hospitalization among uninsured persons associated with the presence of safety net facilities in the county.

Table 3. County-level ACS Hospitalization Rates among Estimated Uninsured Populations, by Age Group, Eight States, 2002.

	Unadjusted Rate per 1000	95% confidence interval	P-value
Children (Ages 0 – 17)			
CHC Only (n = 27)	1.65	(0.98, 2.31)	0.0213
RHC Only (n = 50)	1.12	(0.71, 1.53)	0.3267
RHC&CHC (n = 12)	2.40	(-0.19, 4.99)	0.3973
Neither (n = 160)	1.36	(1.06, 1.67)	n/a
Working age adults (Ages 18 – 64)			
CHC Only (n = 59)	8.44	(7.42, 9.46)	0.0029
RHC Only (n = 137)	11.18	(10.18, 12.17)	0.2361
RHC&CHC (n = 27)	13.30	(9.40, 17.00)	0.1499
Neither (n = 308)	10.40	(9.62, 11.18)	n/a

Source: Authors' analysis using year 2002 State Inpatient Databases representing 8 states, and the 2002 Area Resource File; analysis for children limited to counties having at least 1,000 uninsured children ages 0–17; analysis of adults limited to counties having at least 1,000 uninsured adults ages 18–64.

Table 4. Factors influencing county-level hospital ACS hospitalization rates among estimated uninsured populations, eight states, 2002

	Children (ages 0 – 17) 249 counties			Working age adults (ages 18 – 64) 571 counties		
	Model Coefficient	SE	p-value	Model Coefficient	SE	p-value
Facilities (ref: neither)						
CHC Only	-0.0778	0.2048	0.7042	-0.0089	0.0546	0.8702
RHC Only	-0.1159	0.1606	0.4706	-0.0315	0.0321	0.3266
RHC&CHC	0.4973	0.2125	0.0193	0.0702	0.0579	0.2255
Resources in county:						
MD/DO per 10,000 population	-0.0002	0.0062	0.9780	0.0001	0.0018	0.9340
Beds per 1,000 population	0.0106	0.0265	0.6878	0.0122	0.0045	0.0070
Number of hospitals with Emergency Dept.	0.0130	0.0384	0.7355	0.0227	0.0094	0.0154
HMO penetration rate	-0.0088	0.0045	0.0518	-0.0014	0.0013	0.2761
ED visits per 100	0.0040	0.0028	0.1606	0.0036	0.0005	< .0001
Non-metropolitan county (v metro)	-0.2756	0.1597	0.0843	-0.0549	0.0405	0.1761
Characteristics of county population						
Percent population that is:						
African American	-0.0043	0.0048	0.3724	0.0019	0.0011	0.0840
Hispanic white	-0.0456	0.0280	0.1036	-0.0101	0.0025	< .0001
Asian	0.0710	0.0418	0.0898	-0.0096	0.0161	0.5527
American Indian/Native American	0.2141	0.0390	< .0001	-0.0056	0.0045	0.2193
Population change, 1990 – 2000, %	-0.0216	0.0072	0.0023	-0.0017	0.0014	0.2269
Percent with less than high school education	0.0073	0.0104	0.4787	0.0401	0.0023	< .0001
Population per square mile (/100)	0.0043	0.0013	0.0011	-0.0002	0.0006	0.6908
Percent unemployed (/10)	-0.7131	0.2180	0.0011	0.0884	0.0410	0.0310
Percent uninsured*	-0.0147	0.0139	0.2902	0.0006	0.0052	0.9095
Median household income (thousands)	-0.0241	0.0139	0.0819	-0.0011	0.0013	0.4113
Death rates (×10,000) for:						
Cardiovascular disease	0.0031	0.0093	0.7412	0.0124	0.0023	< .0001
Chronic obstructive pulmonary disease	0.0890	0.0444	0.0447	0.0090	0.0090	0.3194
Diabetes	-0.1353	0.0584	0.0206	0.0194	0.0126	0.1226
Liver disease	-0.3647	0.1294	0.0048	0.0620	0.0230	0.0070

* Estimated percent uninsured among children (ages 0–17) in the model for children, among working age adults (ages 18–64) for the model for adults.

Discussion

Our findings confirm and extend previous research suggesting that CHC presence may be associated with improved access to care, or receipt of care, for certain age groups [25-28]. At the population level, the presence of a CHC in a county was associated with lower ACS admission rates among both working age and older adult populations, when compared to counties that had neither a CHC nor an RHC available. The presence of an RHC in the county was not associated with lower ACS hospitalization rates among children or working age adults. This conflicts with the single previous study exploring RHC effects. However, the work by Zhang and colleagues [30] was restricted to a single type of county (HPSA) in a single state (Nebraska), and was also limited to estimating relative risks of having an ACS diagnosis versus other diagnosis among hospitalized individuals. Their findings may thus be geographically and structurally restricted, and may not reflect risks of ACS hospitalization at the population level across a more diverse region.

Possible Associations Among Older Individuals

Among older adults in the present study, ACS hospitalization rates were lower in counties with CHCs or RHCs, alone or together, compared with counties having neither facility. These rate differences provide suggestive evidence that CHC and RHC location in a county may be associated with greater accessibility or quality of primary health care.

Adjusted admission rates for ACS conditions among older adults were 12% lower among counties that had a CHC plus an RHC, compared with those having no safety net facility, and were 16% lower across counties having only a CHC. The association between CHC presence and lower ACS admission rates may be a function of CHC availability, paralleling earlier research [33], or may be related to chronic disease management programs in CHCs [44]. Further research linking older adults to specific safety net facilities is needed to clarify the findings of the present ecological analysis. The reduction in ACS hospitalization rates among older adults in counties with RHCs was small, but consistent with previous research linking CHC/RHC availability to reduced hospitalization among Medicare beneficiaries. [21]

It is possible that lower ACS hospitalization rates for older individuals in counties having a CHC or RHC are an artifact of differences between counties with and without such facilities, even after adjusting for the factors noted in Table 2. For example, counties with CHCs have markedly higher HMO penetration rates than other counties; higher HMO penetration is associated with lower hospitalization rates. [37]

Minimum Associations Among Children

The presence of a CHC or RHC in the county of residence was associated with ACS hospitalization rates only in counties having both of these facilities, where it was associated with higher rates. These findings contradict previous research suggesting that CHC presence reduced ACS admission rates among children [28]. The study by Garg and associates [28], however, was restricted to a single state (South Carolina) and may not be typical of other U.S. states. The findings of no association between CHC or RHC presence and hospitalization rates are consistent with other research, restricted to urban counties, finding no CHC effects on pediatric hospitalization when physician supply was held constant. [33] The association of CHC plus RHC presence with higher hospitalization rates is unexpected. Further research is needed to ascertain whether this finding was an artifact of the small number of counties studied, or is more generally relevant.

Absence of Effects Among the Uninsured

While the presence of a CHC in a county was associated with lower ACS hospitalization rates at the population level, it did not have parallel associations for the estimated uninsured population. Similarly, the presence of an RHC in a county was not associated with lower estimated hospitalization rates among the uninsured. As noted earlier, RHCs are not required to accept uninsured individuals, and a minority of RHCs report doing so. [8] Thus, it would not be anticipated, on the basis of mission, that RHCs would improve access to care for the uninsured. However, expansion of the number of CHC access points across the nation has been a key element of the Federal approach to the uninsured population since 2002. [45] Absence of CHC effects for the uninsured, assuming that the present ecological study is confirmed by additional research, could indicate the need for a revised approach to improving access.

Further research is needed to clarify individual and institutional barriers to the provision of quality primary care to uninsured populations. Analysts have suggested that CHC expansion has not been sufficient to keep pace with the increasing number of uninsured persons caused by the steady erosion in private insurance. [45] Further, since minorities are more likely than whites to lack insurance, addressing the problem of disparities among the uninsured is key to addressing racial/ethnic disparities in general. [45] Finding measures that will counteract any barriers experienced by uninsured populations will thus contribute to the reduction of race based, as well as insurance based, differences in care.

The present study had several methodological limitations. First, like most studies using the ACS indicator [e.g., [23,33,36]], the analysis was ecological. While the county of residence of hospitalized persons was identified in the SID, no information was available regarding ambulatory care, beyond physician supply and the presence of the types of facility studied. Thus, we are unable to state what proportion of persons in a county received their care from a CHC or RHC, and thus could not directly address the role of these institutions in limiting ACS admissions. Second, the analysis is based on a convenience sample of states providing patient residence data in 2002. An analysis using more recent data for the same states might yield different results, given the expansion in CHC treatment sites and increasing adoption of Health Disparities Collaborative activities in recent years. [44] In addition, the number of states providing residence data to the SIDs has increased; an analysis based on all available states might have different findings. On the other hand, the results of the present analysis are applicable to the large population of the eight states we studied, totaling 72.3 million. Third, the study did not control for the potential presence of additional safety net facilities, such as free clinics. Such facilities could be more likely to locate in counties served by CHCs or RHCs, enhancing the effect of the latter; alternatively, they could be located in other counties and reduce the comparison to study counties. Fourth, the study used estimates of the number of uninsured persons in each county as the denominator for calculating hospitalization rates among the uninsured. While Census estimates offer reasonably accurate estimates of the uninsured population, our results may be limited by measurement error in these estimates. Conversely, insurance information provided in the discharge summary (from which the numerator was calculated) may be inaccurate if information about eventual payor was added at a later point. Fifth, the data did not permit the identification of individuals, and therefore of repeated hospitalizations for the same individual. Repeated hospitalizations for the same individuals may bias the estimated results. Given that the study period was limited to a single year, this factor is unlikely to have affected the results notably. However, it is possible that individuals with ACS hospitalizations are at higher risk of early re-hospitalization due to inadequate follow-up after discharge, given that ACS hospitalizations suggest a problem with the accessibility or quality of primary health care. Sixth, several control variables were obtained from the Area Resource File (ARF), a data source that is commonly used by health services researchers for county-level measures. Supported by the Health Resources and Services Administration, the ARF provides measures for many health-related variables for all U.S. counties. However, some of its measures, such as those from the American Hospital Association, are subject to survey error.

Conclusion

Our results suggest that CHCs and RHCs may play a useful role in providing access to primary health care. Their presence in a county is associated with a lower rate of hospitalization for ambulatory care sensitive conditions among older adults, and in some circumstances for working age adults. Further research is needed to verify to potential relationships suggested by this study, and to understand the role of CHCs and RHCs in access to health care for children.

Competing Interests

The authors declare that they have no competing interests.

Authors' Contributions

JCP developed the study concept, identified applicable data, and drafted the manuscript. JNL designed and implemented the statistical and modeling approach, and helped draft the manuscript. SBL helped to develop the study design and modeling approach, and helped draft the manuscript. All authors read and approved the final manuscript.

Acknowledgements

The research reported here was supported in part by Grant No. 6 U1CRH03711-03 from the Office of Rural Health Policy, Health Resources and Services Administration, US Department of Health and Human Services. We note that the funding agency had no part in the development of this manuscript or its conclusions, for which the authors are solely responsible.

References

1. Economic Research Service, US Department of Agriculture: Rural Poverty at a Glance. [http://www.ers.usda.gov/publications/rdrr100/rdrr100_lowres.pdf]. Rural Development Research Report Number 100 2004.

2. National Center for Health Statistics: Health, United States, 2006. [http://www.ncbi.nlm.nih.gov/books/bv.fcgi?rid=healthus06.TOC]

3. Vlahov D, Galea S: Urbanization, urbanicity, and health. J Urban Health 2002, 79(4 Suppl 1):S1–S12.

4. Bureau of Primary Health Care: Uniform Data System, 2006 National Reports, Table 3b. [http://bphc.hrsa.gov/uds/2006data/National/NationalTable3BUniversal.htm]

5. Bureau of Primary Health Care: Uniform Data System, 2006 National Reports, Table 4. [http://bphc.hrsa.gov/uds/2006data/National/NationalTable4-Universal.htm]

6. Bureau of Primary Health Care: BPHC Policy Information Notice: 98–23. [http://bphc.hrsa.gov/policy/pin9823/default.htm] 1998.

7. Centers for Medicare and Medicaid Services: Rural Health Clinic Fact Sheet. [http://www.cms.hhs.gov/MLNProducts/downloads/rhcfactsheet.pdf]. 2007.

8. Government Accountability Office Health Professional Shortage Areas: Problems Remain with Primary Care Shortage Area Designation System. Washington, DC. GAO-07-84 2006.

9. General Accounting Office: Health Centers and Rural Clinics. Payments Likely to be Constrained Under Medicaid's New System. Washington, DC GAO-01-577 2001.

10. Billings J, Hasselblad V: A Preliminary Study: Use of Small Area Analysis to Assess the Performance of the Outpatient Delivery System of New York City. Report prepared for the Health Systems Agency of New York City. New York, NY 1989.

11. Billings J, Teicholz N: Data Watch: Uninsured patients in District of Columbia hospitals. Health Affairs 1990, 9:158–165.

12. Billings J: Consideration of the use of small area analysis as a tool to evaluate barriers to access. In Health Resources and Services Administration, Consensus on Small Area Analysis. DHHS Pub. No. HRSA-PE 91-1[A]. Washington, D.C.: U.S. Department of Health and Human Services; 1990.

13. Billings J, Zeitel L, Lukomnik J, Carey TS, Blank AE, Newman L: Impact of Socioeconomic Status On Hospital Resource Use in New York City. Health Affairs 1993, 12:162–73.

14. Billings J, Anderson GM, Newman LS: Recent findings on preventable hospitalizations. Health Affairs 1996, 15:239–249.

15. Bindman AB, Grumbach K, Osmond D, Komaromy M, Vranizan K, Lurie N, Billings J, Stewart A: Preventable Hospitalizations And Access To Health Care. Journal of the American Medical Association 1995, 274:305–11.

16. Weissman JS, Gatsonis C, Epstein AM: Rates of Avoidable Hospitalization By Insurance Status in Massachusetts and Maryland. Journal of the American Medical Association 1992, 268:2388–94.

17. Pappas G, Hadden WC, Kozak LJ, Fisher GF: Potentially Avoidable Hospitalizations: Inequalities in Rates Between US Socioeconomic Groups. American Journal of Public Health 1997, 87:811–6.

18. Laditka JN: Physician Supply, Physician Diversity, and Outcomes of Primary Health Care for Older Persons in the United States. Health and Place 2004, 10:231–44.

19. Ansari Z, Laditka JN, Laditka SB: Access to Health Care And Hospitalization For Ambulatory Care Sensitive Conditions. Medical Care Research and Review 2006, 63:719–41.

20. Silver MP, Babitz ME, Magill MK: Ambulatory Care Sensitive Hospitalization Rates In The Aged Medicare Population In Utah, 1990 To 1994: A Rural-Urban Comparison. Journal of Rural Health 1997, 13:285–94.

21. Culler SD, Parchman ML, Przybylski M: Factors Related To Potentially Preventable Hospitalizations Among The Elderly. Medical Care 1998, 36:804–17.

22. Laditka JN, Laditka SB, Probst JC: Hospitalization for Ambulatory Care Sensitive Conditions across Levels of Rurality. Health and Place 2009, 15:731–40.

23. DeLia D: Distributional Issues In The Analysis Of Preventable Hospitalizations. Health Services Research 2003, 38(6 Pt 2):1761–79.

24. Agency for Healthcare Research and Quality: Monitoring the Healthcare Safety Net. [http://www.ahrq.gov/data/safetynet/databooks/safetynet_key1.htm]

25. Falik M, Needleman J, Wells BL, Korb J: Ambulatory Care Sensitive Hospitalizations And Emergency Visits: Experiences Of Medicaid Patients Using Federally Qualified Health Centers. Medical Care 2001, 39:551–61.

26. Falik M, Needleman J, Herbert R, Wells B, Politzer R, Benedict MB: Comparative Effectiveness Of Health Centers As Regular Source Of Care: Application Of Sentinel ACSC Events As Performance Measures. Journal of Ambulatory Care Management 2006, 29:24–35.

27. Epstein AJ: The Role Of Public Clinics In Preventable Hospitalizations Among Vulnerable Populations. Health Services Research 2001, 36:405–20.

28. Garg A, Probst JC, Sease T, Samuels ME: Potentially Preventable Care: Ambulatory Care-Sensitive Pediatric Hospitalizations In South Carolina In 1998. Southern Medical Journal 2003, 96:850–8.

29. Schoenman JA, Cheng CM, Evans WN, Blanchfield BB, Mueller CD: Do Hospital-Based Rural Health Clinics Improve The Performance Of The Parent Hospital? Policy Analysis Brief W series/Project Hope, Walsh Center for Rural Health Analysis 1999, 2:1–4.

30. Zhang W, Mueller KJ, Chen LW, Conway K: The Role Of Rural Health Clinics In Hospitalization Due To Ambulatory Care Sensitive Conditions: A Study In Nebraska. Journal of Rural Health 2006, 22:220–3.

31. Agency for Healthcare Research and Quality: Safety Net Monitoring, Appendix B. [http://www.ahrq.gov/data/safetynet/billappb.htm] Ambulatory Care Sensitive Conditions 2007.

32. Andersen RM: Revisiting the Behavior Model and Access to Medical Care: Does it Matter? Journal of Health and Social Behavior 1995, 36:1–10.

33. Laditka JN, Laditka SB, Probst JC: More May be Better: Evidence of a Negative Relationship between Physician Supply and Hospitalization for Ambulatory Care Sensitive Conditions. Health Services Research 2005, 40:1148–66.

34. Schreiber S, Zielinski T: The meaning of ambulatory care sensitive admissions: Urban and rural perspectives. Journal of Rural Health 1997, 13:276–284.

35. Krakauer H, Jacoby I, Millman M, Lukomnik JE: Physician impact on hospital admission and on mortality rates in the Medicare population. Health Services Research 1996, 31:191–211.

36. Ricketts TC, Randolph R, Howard HA, Pathman D, Carey T: Hospitalization rates as indicators of access to primary care. Health and Place 2001, 7:27–38.

37. Zhan C, Miller MR, Wong H, Meyer GS: The effects of HMO penetration on preventable hospitalizations. Health Services Research 2004, 39:345–61.

38. Bindman AB, Chattopadhyay A, Osmond DH, Huen W, Bacchetti P: The impact of Medicaid managed care on hospitalizations for ambulatory care sensitive conditions. Health Services Research 2005, 40:19–38.

39. US Census Bureau: Metropolitan and Micropolitan Statistical Areas. [http://www.census.gov/population/www/metroareas/metrodef.html]

40. Laditka JN, Laditka SB, Mastanduno MP: Hospital Utilization for Ambulatory Care Sensitive Conditions: Health Outcome Disparities Associated with Race and Ethnicity. Social Science and Medicine 2003, 57:1429–1441.

41. Laditka JN, Laditka SB: Race, Ethnicity, and Hospitalization for Six Chronic Ambulatory Care Sensitive Conditions in the United States. Ethnicity and Health 2006, 11:247–263.

42. Laditka JN: Hazards of Hospitalization for Ambulatory Care Sensitive Conditions among Older Women: Evidence of Greater Risks for African Americans and Hispanics. Medical Care Research and Review 2003, 60:468–495.

43. United States Census Bureau: Small Area Health Insurance Estimates: Model-based Estimates for Counties and States. [http://www.census.gov/did/www/sahie/data/index.html] U.S. Census Bureau, Data Integration Division, Small Area Estimates Branch: Washington, DC;

44. Health Disparities Collaboratives: Background. [http://www.healthdisparities.net/hdc/html/about.background.aspx]

45. Hadley J, Cunningham P, Hargraves JL: Would Safety-Net Expansions Offset Reduced Access Resulting From Lost Insurance Coverage? Race/Ethnicity Differences. Health Affairs 2006, 25:1679–87.

Harm Reduction in Hospitals: Is it Time?

Beth S. Rachlis, Thomas Kerr, Julio S. G. Montaner
and Evan Wood

ABSTRACT

Among persons who inject drugs (IDU), illicit drug use often occurs in hospitals and contributes to patient expulsion and/or high rates of leaving against medical advice (AMA) when withdrawal is inadequately managed. Resultant disruptions in medical care may increase the likelihood of several harms including drug resistance to antibiotics as well as costly readmissions and increased patient morbidity. In this context, there remains a clear need for the evaluation of harm reduction strategies versus abstinence-based strategies with respect to addressing ongoing issues related to substance use among addicted hospitalized patients. While hospitalization can be used to stabilize addicted patients as they recover from their acute illness and help them to achieve abstinence, patients unable to maintain abstinence should not be penalized for failing to do so at the expense of their health. This article describes harm reduction activities within hospitals and areas for future investigation.

Introduction

Soft-tissue infections and other injection-related infections are among the main contributors to health service use among people who inject drugs (IDU) [1-6]. In many settings, the two most common reasons for emergency department (ED) visits relate to soft-tissue infections, and problems related directly to drug use (e.g., overdose)[1,2,4,6]. Not-surprisingly, many IDU use EDs as a regular point of care; IDU are generally less likely to use outpatient services compared to non-IDU[4] and generally face poor access to prevention programs and addiction treatment services [7-9].

As a result, IDU often present to EDs later in the course of their illness, and this in turn increases the likelihood for hospital admission [2,4,5]. Drug-related infections are often painful and may progress to more serious life- and limb-threatening conditions [10]. More complicated infections such as endocarditis require extended periods of treatment with intravenous antibiotics and thus may require even longer hospital stays.

However, IDU are more likely than other patients to discharge from hospitals against medical advice (AMA) [11,12]. A 2002 study noted that IDU were over four times more likely to leave AMA compared to non-IDU [12] and leaving AMA is a strong predictor for frequent readmission [11-13]; Moreover, repeated admissions for chronic medical problems are generally more costly for total days of stay than single, cost-intensive stays [13].

In addition to the high costs associated with increased health utilization, these findings also suggest that patients are not fully recovering from their illness the first time they are treated. Incomplete therapy or treatment failure may also increase the likelihood for drug resistance to antibiotics [11,13,14]. As such, uncovering why IDU are more likely to leave AMA is a necessary first step in order to improve health outcomes, although incidentally this may also decrease the high costs associated with elevated rates of health service utilization.

Discussion

Harm Reduction

While an abstinence-based approach to drug use generally requires that complete cessation from all non-prescribed drugs is a pre-requisite for effective addiction treatment [15], harm reduction emphasizes that efforts to improve health and social outcomes should begin with 'where a person is at' in terms of their drug use [16]. Strategies need to be maximized, both in terms of types of services offered and where they operate. Furthermore, abstinence-based programs are generally considered high-threshold referring to the eligibility criteria for participation in such programs and the state of 'readiness' individuals need to be in prior to entry [16,17]. Low threshold services, including needle exchange programmes (NEPs), have minimal requirements for involvement and put IDU in contact with a continuum of care even when they may not be ready to engage in abstinence-based treatment [18]. Harm reduction involves a continuous spectrum of strategies, from the promotion of safer and managed drug use to complete abstinence [15]. Harm reduction advocates and guidelines [19] suggest that strategies to reduce the high risk of disease transmission should be culturally relevant and implemented within multiple contexts, including health care facilities such as hospitals [18]. Indeed, evidence suggests that active drug use does occur in hospitals and is associated with leaving AMA [12,20].

In terms of specific strategies, methadone maintenance treatment (MMT) has been associated with reductions in the need for hospitalization and generally results in improvements in health care access [2,20]. NEPs work to reduce disease transmission by lowering the rate of syringe sharing and the number and length of time used syringes are in circulation [7,21-24]. Supervised Injecting Facilities (SIFs) have also demonstrated success in the reduction of HIV risk and other harms among IDU. At North America's first SIF, IDU are provided with sterile syringes, primary care services, and referral to addiction treatment, as well as to emergency care [25]. SIF use has been associated with increases in safer injecting practices [26,27], more rapid entry into detox programs [27] and generally increased uptake of addiction treatment [9].

Gaps in Service Delivery

While achieving abstinence from illicit drug use is ideal, for many individuals, this may be difficult, particularly without adequate support. Health care for drug users often follows psychiatric models of care that involve the use of contracts developed for addiction management. When this contract is breached (i.e., drug use continues), the patient may be discharged back into the community with cessation of care [28]. Such approaches have potentially significant ethical implications as they may impede appropriate care for drug users [29].

Negative experiences with the medical establishment may also impede health care delivery for IDU [30]. Leaving hospital AMA predisposes individuals not only to poor health outcomes due to inadequate treatment but also to major disruptions in the patient-provider relationship [20]. Recently, our local teaching hospital generated controversy when a strict illicit drug use policy that essentially allows for 'evictions' of drug users who are unable to maintain abstinence while in hospital was proposed. While this policy is currently under review, similar guidelines are in place in most hospitals in North America. The fact that active drug use occurs in hospitals and is one reason why many IDU leave AMA raises the question that if active drug use was accommodated rather than banned in hospitals, rates of leaving AMA would decline. While incorporating harm reduction in hospitals to deal with addicted patients raises a host of ethical and well as staff and patient safety issues, such an approach has the potential to result not only in better health outcomes but reduced readmissions.

Incorporation of Harm Reduction Programs

Indeed, harm reduction programs have already shown success when integrated with medical care. Increased integration of low- and medium-threshold harm reduction strategies with primary and acute care has been associated with increasing the proportion of IDU who have regular health care [28].

For instance, the Dr. Peter Centre in Vancouver which provides low-threshold access to care for people living with HIV/AIDS including a high proportion of IDU offers one example where harm reduction has been successfully integrated with a medical facility. Many conventional barriers have been removed at the Centre including the need to remain drug-free. MMT and the distribution of condoms and clean needles are also provided [30]. An interdisciplinary team embraces harm reduction through the promotion of self-care and autonomy and in the spring of 2002, the nurses implemented a pilot project involving the supervision of injections in the nursing treatment room. An opiate-overdose protocol was also developed and illicit drugs including crack cocaine can be smoked in a designated area on the premises. By May 2003, staff had noted a reduced incidence of soft-tissue infections associated with use of the injecting room [31].

At the Dr. Peter Centre, participants are able to build trusting relationships with healthcare staff; such a facility offers an important solution to increase acceptability of care while reducing stigma among IDU. Importantly, the continuity of care from both nurses and doctors has shown to be an effective means for reducing injection-related complications and the need for hospital admission [28,30].

Specific harm reduction strategies including drug substitution for opioid addiction, smoking rooms for tobacco and illicit drugs, and protocols to help manage drug withdrawal symptoms have already demonstrated success in their integration into health care facilities and should continue to be fully implemented into hospitals. For example, in-patient MMT has been associated with a reduced likelihood of leaving AMA which may reflect adequate and appropriate management for opioid withdrawal [20]. Certifying a greater number of physicians who are able to prescribe buprenorphine has also already been shown to result in a reduced number of hospitalizations and risk of complications [32]. Providing patients presenting with obvious physical withdrawal with additional doses of opiates or short courses of benzodiazepines has been associated with reductions in agitation and early discharge [20].

Other strategies, while proven effective in community settings, still require further study given their potential role in reducing harm among hospital-admitted IDU. Supervised injecting areas and NEPs, in particular, could be evaluated as services that could be made accessible for hospital patients, particularly those with longer stays or in wards that are designated for dealing with addicted individuals. Ideally, the availability of these services would also help to facilitate positive patient-provider relationships.

Conclusion

Active drug use occurs in hospitals and contributes to high rates of leaving AMA among IDU. As discussed, if active drug use was accommodated through more of a harm reduction approach rather than banned in hospitals, rates of leaving AMA would likely decline. Regardless, there remains the need for evaluation of several novel harm reduction interventions versus abstinence-based strategies with respect to addressing ongoing issues related to stigmatization and elevated rates of leaving AMA. This may lend itself to a randomized trial or perhaps it is better examined via observational data where the objective would be to evaluate whether the incorporation of a different harm reduction programs (e.g., safer injecting spaces) in hospitals results in reduced rates of patients leaving AMA and overall improvements in health outcomes for IDU who are able to access these services versus those who do not. Given the contact that many IDU have with EDs, it seems fitting that harm reduction programs should continue to expand to the hospital setting, particularly when the number of IDU being treated is high. The goal is to use hospitalization to stabilize addicted patients as they recover from their acute illness and see if they can be helped to achieve abstinence. However, patients unable to maintain abstinence should not be penalized for failing to do so at the expense of their health.

Competing Interests

The authors declare that they have no competing interests.

Authors' Contributions

EW and BR developed the concept of the manuscript. BR drafted the original version. TK, EW, and JSG assisted with revisions. All authors approved the final manuscript.

References

1. Binswanger IA, Kral AH, Bluthenthal RN, Rybold DJ, Edlin BR: High prevalence of abscesses and cellulitis among community-recruited injection drug users in San Francisco. Clin Infect Dis 2000, 30:579–81.

2. Palepu A, Tyndall MW, Leon H, Muller J, O'Shaughnessy MV, Schechter MT, Anis A: Hospital utilization and costs in a cohort of injection drug users. CMAJ 2001, 165:415–20.

3. Stein MD, Anderson B: Injection frequency mediates health service use among persons with a history of drug injection. Drug Alcohol Depend 2003, 70:159–68.

4. French MT, McGreary KA, Chitwood DD, McCoy CB: Chronic illicit drug use, health services utilization and the cost of medical care. Soc Sci Med 2000, 50:1703–13.

5. Kerr T, Wood E, Grafstein E, Ishida T, Shannon K, Lai C, Montaner J, Tyndall MW: High rates of primary care and emergency department use among injection drug users in Vancouver. J Public Health 2005, 27:62–66.

6. Lloyd-Smith E, Kerr T, Hogg RS, Li K, Montaner JS, Wood E: Prevalence and correlates of abscesses among a cohort of injection drug users. Harm Reduct J 2005, 2:24.

7. Stathdee SA, Patrick DM, Currie SL, Cornelisse PGA, Rekart ML, Montaner JSG, Schechter MT, O'Shaughnessy MV: Needle exchange is not enough: lessons from the Vancouver Injecting drug users study. AIDS 1997, 11:f59–65.

8. Vlahov D, Celentano DD: Access to highly active antiretroviral therapy for injection drug users: adherence, resistance, and death. Cad Saude Publica 2006, 22:705–18.

9. Wood E, Li K, Palepu A, Marsh D, Schechter MT, Hogg R, Montaner J, Kerr T: Sociodemographic disparities in access to addiction treatment among a cohort of injection of Vancouver injection drug users. Subst Use Misuse 2005, 40:1153–67.

10. Takahashi TA, Baernstien A, Binswanger I, Bradley K, Merrill JO: Predictors of hospitalization for injection drug users seeking care for soft tissue infections. JGIM 2007, 22:382–388.

11. Jeremiah J, O'Sullivan P, Stein MD: Who leaves against medical advice? JGIM 1995, 10:403–5.

12. Anis AH, Sun H, Guh DP, Palepu A, Schechter MT, O'Shaughnessy MV: Leaving hospital against medical advice among HIV-positive patients. CMAJ 2002, 167:633–7.

13. Palepu A, Sun H, Kuyper L, Schechter MT, O'Shaughnessy MV, Anis AH: Predictors of early hospital readmission in HIV-infected patients with pneumonia. JGIM 2003, 18:242–247.

14. Whynot E: Health impact of injection drug users and HIV in Vancouver. Vancouver Health Board; 1996.

15. Peterson J, Gwin Mitchell S, Hong Y, Agar M, Latkin C: Getting clean and harm reduction: adversarial or complementary issues for injecting drug users. Cad Saude Publica 2006, 22:733–40.

16. Lenton S, Single E: The definition of harm reduction. Drug Alcohol Rev 1998, 17:213–9.

17. Kerr T, Palepu A: Safe injection facilities in Canada: is it time? CMAJ 2001, 165:436–7.

18. Marlatt GA: Harm reduction: come as you are. Addict Behav 1996, 21:779–88.

19. World Health Organization: Policy and programming guide for HIV/AIDS prevention and care among injection drug users. [http://www.undoc.org/documents/hiv-aids/policy%20programming%20guide.pdf]. WHO; 2005.

20. Chan ACH, Palepu A, Guh DP, Sun H, Schechter MT, O'Shaughnessy MV, Anis AH: HIV-positive injection drug users who leave the hospitals against medical advice. The mitigating role of methadone and social support. JAIDS 2004, 35:56–59.

21. Hurley SF, Jolley DF: Effectiveness of needle-exchange programmes for prevention of HIV infection. Lancet 1997, 349:1797–1801.

22. Blutenthal RN, Kral AH, Gee L, Erringer E, Edlin BR: The effect of syringe exchange use on high-risk injection drug users: a cohort study. AIDS 2000, 14:605–11.

23. Des Jarlais DC, Marmor M, Paone D, Titus S, Shi Q, Perlis T, Jose B, Friedman SR: HIV incidence among injecting drug users in New York City syringe-exchange programmes. Lancet 1996, 348:987–91.

24. Dolan K, Kimber J, Fry C, Fitzgerald J, McDonald D, Trautmann F: Drug consumption facilities in Europe and the establishment of supervised injecting centres in Australia. Drug Alcohol Rev 2000, 19:337–46.

25. Kerr T, Tyndall M, Li K, Montaner J, Wood E: Safer injection facility use and syringe sharing in injection drug users. Lancet 2005, 366:316–18.

26. Wood E, Tyndall M, Stolz J, et al.: Safer injecting education for HIV prevention within a medically supervised safer injecting facility. Int J Drug Policy 2005, 16:281–4.

27. Wood E, Tyndall MW, Zhang R, Stolz JA, Lai C, Montaner JS, Kerr T: Attendence at supervised injecting facilities and use of detoxification services. NEJM 2006, 354:2512–4.

28. Heller D, McCoy , Cunningham C: An invisible barrier to integrating HIV primary care with harm reduction services: philosophical clashes between the harm reduction and medical models. Pub Health Reps 2004, 119:32–39.

29. Werb D, Elliot E, Fischer B, Wood E, Montaner J, Kerr T: Drug treatment courts in Canada: an evidence-based review. HIV/AIDS Policy Law Rev 2007, 12:12–17.

30. Griffiths H: Dr. Peter Centre: removing barriers to health care services. Nurs BC 2002, 34:10.

31. Wood A, Zettel P, Stewart WIL: Harm reduction nursing: The Dr. Peter Centre. Can Nurs 2003, 99:20.

32. Jacobsohn V, DeArman M, Moran P, Cross J, Dietz D, Allen R, Bachofer S, Dow-Velarde L, Kaufman A: Changing hospital policy from the wards: an introduction to health policy education. Acad Med 2008, 83:352-56.

"Good Idea but not Feasible" — The Views of Decision Makers and Stakeholders Towards Strategies for Better Palliative Care in Germany: A Representative Survey

Sara Lena Lueckmann, Mareike Behmann,
Susanne Bisson and Nils Schneider

ABSTRACT

Background

*Statements on potential measures to improve palliative care in Germany pre-
dominantly reflect the points of view of experts from specialized palliative care*

organizations. By contrast, relatively little is known about the views of representatives of organizations and institutions that do not explicitly specialize in palliative care, but are involved to a relevant extent in the decision-making and policy-making processes. Therefore, for the first time in Germany, we carried out a representative study of the attitudes of a broad range of different stakeholders acting at the national or state level of the health care system.

Methods

442 organizations and institutions were included and grouped as follows: patient organizations, nursing organizations, medical associations, specialized palliative care organizations, political institutions, health insurance funds and others. Using a standardized questionnaire, the participants were asked to rate their agreement with the World Health Organization's definition of palliative care (five-point scale: 1 = completely agree, 5 = completely disagree) and to evaluate 18 pre-selected improvement measures with regard to their general meaningfulness and the feasibility of their introduction into the German health care system (two-point scale: 1 = good, 2 = poor).

Results

The response rate was 67%. Overall, the acceptance of the aims of palliative care in the WHO definition was strong. However, the level of agreement among health insurance funds' representatives was significantly less than that among representatives of the palliative care organizations. All the improvement measures selected for evaluation were rated significantly higher in respect of their meaningfulness than of their feasibility in Germany. In detail, the meaningfulness of 16 measures was evaluated positively (70–100% participants chose the answer "good"); for six of these measures feasibility was evaluated negatively (0–30% "good"), while for the remaining ten measures feasibility was evaluated inconsistently (31–69% "good").

Conclusion

The reason why potentially meaningful improvement measures are considered to be not very feasible in Germany may be the existence of barriers resulting from the high degree of fragmentation of health care provision and responsibilities. In overcoming these barriers and further improving palliative care it may be helpful that the basic understanding of the palliative care approach seems to be quite homogenous among the different groups.

Background

Palliative care is an approach that improves the quality of life of patients and their families facing the problems associated with life-threatening illness [1]. Despite some advances in recent years, palliative care is still an underdeveloped field in many countries with health systems that are in other respects highly developed [2-5].

In Germany there is a widespread undersupply of specialist palliative care both for inpatients and in particular for outpatients. At present, there are about 330 hospices and palliative care units nationwide, with a total of about 2,800 beds (i.e. about 34 beds per million inhabitants), and about 60 palliative care teams in all (for 80 million inhabitants). Moreover, there are considerable regional differences, with very well developed services in some urban areas and extensive gaps in the periphery [6-8]. Besides, there are problematical deficits in the education and the advanced training of all health care professions regarding palliative care; for example, the subject of palliative medicine is not compulsory in medical schools, and is therefore only taught here and there [9].

However, in recent years palliative care has achieved increasing recognition among the public and in political circles. For example, a big step forming part of the recent (2007) health care reform was the legal introduction of specialist outpatient palliative care (SAPV), i.e. outpatient care delivered by palliative care teams [10]. So far, the realization of SAPV is very problematic; e.g., there are no uniform standards regarding the staffing and structural requirements of palliative care teams, and their cooperation with other players (especially family doctors, nursing services, specialist physicians in other medical disciplines) is not consistently regulated [11].

It has become clear that the further development of palliative care is a major social and political task [6,7,9,12]. In this context it is important to consider that the political decision-making processes in Germany are influenced by numerous lobby groups, organizations and institutions, due to the legal principles of federalism, the self-governing status of medical services and health insurance funds and the separation between outpatient and inpatient services [13,14]. Evaluating the attitudes of the different stakeholders is therefore an important aim of health service research.

So far, statements on potential measures to improve palliative care have mainly reflected the points of view of experts from specialized palliative care and hospice organizations and institutions. By contrast, very little is known about the views of representatives of organizations and institutions that do not explicitly specialize in palliative care, but are involved to a relevant extent in the decision-making and policy-making processes [15]. These are, for example, the various medical and nursing associations, health insurance funds, patients' organizations, and government and regulatory authorities and political parties (these last mentioned being referred to collectively below as "political institutions").

For the first time in Germany we wanted to representatively evaluate the views of a wide range of different stakeholders, acting at the meso and macro level of the health system, concerning selected improvement measures for palliative care. The study was part of a larger research project aimed at developing public health targets for palliative care in Germany. The conceptual framework and design of the overall project are published elsewhere [16].

Methods

Ethics Approval

The Ethics Committee of Hannover Medical School approved the study (letter of 26/02/07).

Sample

For the recruitment of the study population a snowball sampling approach was used. First of all, the members of the study group compiled a list of institutions and organizations based on their respective individual experience (e.g. NSCH is a lecturer in public health and health system research as well as a consultant in family medicine and palliative medicine). The list was supplemented out of the literature and by internet research, as well as by discussion rounds with further public health specialists in our department.

The main inclusion criteria for the organizations were:

* being involved in the provision, financing and/or political organization of health care;
* being associated with a lobby group of health care professionals, patients, politicians and/or health insurance funds;
* being active at the state or national level of the German health care system.

This led to our including of 366 institutions and organizations in the first round of the survey. In order to identify further relevant players, we asked the participants initially included: "An important aim of this study is to assess the views of all relevant stakeholders within the context of palliative and end-of-life care in Germany. We would therefore like to ask you to write down the names of other players, institutions and organizations which in your opinion should also be questioned." In this way we identified a further 76 stakeholders that had not already been included in the first round.

In the end, 442 organizations and institutions were included and grouped (Table 1). The groups had been predefined when the project was designed [16]: A: patient organizations, B: medical associations, C: nursing organizations, D: health insurance funds, E: political institutions, F: specialized palliative care organizations and G: others. The group "others" is included in the outcome of analyses for all groups, but not in the comparison of the groups because of its high level of heterogeneity.

Table 1. Group sizes and characteristics of organizations

ID	Group	group size sample (contacted)	response rate	final group size (analyzed)	proportion (of all analyzed)	characteristics (examples)
	all groups	442	67%	301	100%	
A	patient organizations	42	57%	24	8%	umbrella organizations of self-help groups and senior citizens interest groups, federation of German consumer centres
B	medical associations	100	78%	79	26%	scientific medical societies, associations of statutory health insurance physicians on federal and national level
C	nursing organizations	22	77%	17	6%	umbrella organizations of nursing care, federal working committee of nursing
D	health insurance funds	82	63%	55	18%	associations of health insurance funds on federal and national level, umbrella organizations of health insurance funds
E	political institutions	45	49%	24	8%	state and federal ministries of health, health-care policy spokespeople of the different political parties
F	specialized palliative care	54	65%	39	13%	associations and umbrella organizations of palliative and hospice care on federal and national level
G	others*	97	72%	63	21%	charity and clerical institutions, associations for physiotherapy, federal joint committee

* not included in the comparison of the groups because of the strong heterogeneity

Procedure

The 366 institutions approached in the first round were sent a letter of advice in April 2008 in which they were informed about the study and invited to participate. Initially, the letter was addressed to the heads of the institutions and organizations. The addressees were asked to forward the questionnaire to another person within the organization if they themselves did not feel responsible or qualified enough in palliative care.

One week after the information letter we mailed the questionnaires together with further information, instructions and a stamped addressed envelope. Four weeks later all organizations received a postcard of thanks which included the request to fill in the questionnaire if they had not already done so. Another five weeks later, those organizations that had not yet answered received a further reminder with one more questionnaire.

The additional 76 stakeholders identified later were addressed in the same way. The survey was carried out from April to August 2008.

Instrument

We developed a standardized questionnaire. To study the participants' basic understanding of the palliative care approach, we sought to identify the acceptance of selected aspects from the World Health Organization's definition of palliative care [1] using a five-point scale (1 = completely agree, 2 = agree, 3 = undecided, 4 = disagree, 5 = completely disagree).

Additionally, the questionnaire contained 18 items that presented potential measures aiming to improve palliative care in Germany. These measures were selected on the basis of literature reviews and of our own pre-studies [e.g. [15]]. Their potential relevance to the development of public health targets for palliative care in Germany [16] was also taken in consideration. The participants were asked to assess the measures with regard to their meaningfulness in general and also to their feasibility in Germany. The answers could be given using a two-point scale (1 = good, 2 = poor).

In order to evaluate the comprehensibility and practicability of the questionnaire, a cognitive pre-test was carried out with 13 participants not included in the study. The pre-test participants had to fill in the questionnaire and reply to additional questions focusing on their understanding of the questionnaire (probing method). Furthermore, the participants could ask about the tenor of the questions. After seven pre-tests we interrupted the testing since we detected some room for improvement. After revising the questionnaire we carried out another six pre-tests resulting in some further minor revisions.

Statistics

Data were recorded with MS Access (2003) and transferred to SPSS for Windows 16.0 for statistical analysis. Data are reported descriptively as frequencies in percent. The McNemar test was used to analyze the differences between the evaluations of the meaningfulness and the feasibility of the improvement measures; the chi-square test was used to examine proportional differences between the groups.

The evaluation of the World Health Organization's definition of palliative care on a five-point scale was reduced to a three-point scale to avoid too small cell frequencies. Because of the asymmetrical distribution we decided to combine the answer choices 3–5 (3 = undecided, 4 = disagree, 5 = completely disagree) into one (3 = undecided, disagree and completely disagree).

We defined the assessment of measures as positive when = 70% of participants chose the answer "good," and as negative when = 30% chose the answer "good." The answers were assessed as being inconsistent when 31–69% of participants chose the answer "good." $P < .05$ was considered statistically significant.

Results

295 stakeholders replied to the questionnaire. Overall, the response rate was 67%. The response rate was best for the group of medical associations (78%, n = 78) and worst for the group of political institutions (49%, n = 22). Three organizations asked for more than one questionnaire or copied the original because representatives from different departments within the organization wanted to participate. In these three exceptional cases we accepted this. As a result, the group size was changed for three groups: medical associations (n+1), health insurance funds (n+3) and political institutions (n+2). In the end, we were able to analyze a total of 301 questionnaires (Table 1).

The demographic data of the respondents are shown in Table 2. The respondents were between 25 and 73 years of age (mean 51), 62% were male and 40.7% had studied medicine. 26.8% of the questionnaires were answered by a member of the board and 24.4% by a head of department. 53% of the participants had been employed in the institutions for more than 10 years.

Table 2. Characteristics of the respondents

demographic data	%	characteristics
age		51 ± 9.0 (Range: 25–73 years)
	26.8	till 45 (n = 75)
	45.0	46–55 (n = 126)
	28.2	56+ (n = 79)
sex	62.0	male (n = 184)
	38.0	female (n = 114)
nationality	99.0	German (n = 293)
professional background*	40.7	medicine (n = 120)
	12.2	economics (n = 36)
	9.8	social science (n = 29)
	13.2	nursing (n = 39)
	8.5	others (n = 25)
position within the organization/institution*	26.8	member of the board (n = 79)
	24.4	head of department (n = 72)
	19.7	business manager (n = 58)
	16.6	consultant (n = 49)
	4.7	volunteer (n = 14)
time of employment in the institution	53.0	> 10 years (n = 157)
	22.3	6–10 years (n = 66)
	22.0	1–5 years (n = 65)
	2.7	<1 year (n = 8)
religion	40.6	catholic (n = 119)
	34.8	protestant (n = 102)
	23.2	undenominational (n = 68)
	1.4	other religion (n = 4)

* multiple answers possible

The groups differed significantly in distribution of sex (p = .004): in the groups patient organizations (A), nursing organizations (C), political institutions (E) and specialized palliative care (F) there were more women, whereas in the groups medical associations (B) and health insurance funds (D) there were more men answering the questionnaire. There was no significant difference between the groups in the distribution of age (p = .12).

Understanding of Palliative Care

The acceptance of the seven aspects of the World Health Organization's definition of palliative care that were asked about was strong. The most pronounced agreement was found to the statement "In my opinion it is part of appropriate palliative care to provide relief from pain and other distressing symptoms": 99% (n = 298) of the respondents showed complete or predominant agreement. Relatively less pronounced agreement was found to the statement "In my opinion it is part of appropriate palliative care to intend neither to hasten nor to postpone death": 72% (n = 213) of the respondents showed complete or predominant agreement.

Comparing the six groups with regard to their weighting of the aspects of the World Health Organization's definition we found statistically significant differences with regard to six out of seven statements. By way of example, the statement "... to integrate the psychological and spiritual aspects of patient care" was most positively assessed by the representatives of specialized palliative care organizations (97.4% agreed completely) and least positively by the health insurance funds' representatives, of whom only 49.1% completely agreed. The statement "... to offer a support system to help the family cope during the patient's illness" was most positively assessed by the representatives of specialized palliative care organizations (79.5% agreed completely) and least positively by the health insurance funds' representatives (of whom only 20% completely agreed).

All in all, in contrast to the specialized palliative care organizations' representatives, the health insurance funds' representatives agreed considerably less with the WHO definition of palliative care.

Assessment of Improvement Measures

Overall, the feasibility of potential improvement measures in Germany was evaluated considerably less positively than their meaningfulness. The discrepancy between meaningfulness and feasibility is significant for all 18 measures that were presented (p < .001).

In detail, the meaningfulness of 16 measures was evaluated as being positive (= 70% "good"), whereas the feasibility of six out of these 16 measures was evaluated as being negative (= 30% "good"). This concerns all measures regarding the topic caring time (items 5–8) as well as the measures of nursing home physicians (item 14) and academic centres for research (item 17). For these six measures the discrepancy is the greatest. The feasibility of the remaining ten (out of these 16) measures was evaluated inconsistently (31–69% "good").

Both the meaningfulness and the feasibility of the measure "Palliative care patients are completely exempted from additional payments for drugs, cures and aids" (item 12) were evaluated inconsistently.

In the case of the measure "Case managers organize and coordinate the care of palliative care patients (item 15)," was the meaningfulness evaluated inconsistently while the feasibility was evaluated negatively.

Comparing the six groups with regard to their rating of the meaningfulness and feasibility of the measures we found statistically significant differences for some statements:

- Representatives from patient organizations and/or specialized palliative care evaluated the meaningfulness of six measures (items 2, 3, 8, 10, 11, 17), namely those concerning the topics education and training, as well as availability of 24-7 regional palliative care services and establishment of publicly funded academic centres for research in palliative care, more highly than the other groups did.
- Nursing organizations evaluated the meaningfulness of three measures more highly than the other groups did: palliative care patients are completely exempted from additional payments for drugs, cures and aids (item 12), regular availability of nursing home physicians with specialized training in palliative care (item 14), case managers organize and coordinate the care of palliative care patients (item 15).
- Health insurance funds evaluated the feasibility of compulsory training in palliative medicine for medical students (item 2) more highly than the other groups did.

To sum up, the representatives of patient, nursing and specialized palliative care organizations evaluated the meaningfulness of the proposed improvement measures for palliative care in Germany considerably better than the other groups did. However, the feasibility of all improvement measures received a significantly worse assessment in all cases compared to the assessment of their meaningfulness.

Discussion

This study is, to our knowledge, the first one that gives representative insights into the views of a broad range of decision-makers and lobby groups at the meso and macro level of the German health system with regard to various improvements in palliative care. The results broaden the perspectives and point to possible strategies aimed at promoting further scientific and political development in the field. For example, they might be integrated into the "charter process for the care of very severely ill and dying people" that has recently been initiated by the German Association for Palliative Medicine (DGP), the German Hospice and Palliative Organization (DHPV) and the German Medical Association (BÄK) [17]. The aims of this charter are to promote dialogue between all those concerned and the involvement of society with the subject, to provide guidance for the direction of future developments and to reach agreement on common objectives and actions. The charter relates above all to issues of social policy, especially to ethical and legal issues, the further development of care structures, questions of initial and in-service training in the various professions concerned and issues relating to research.

The response rate (67%) was good considering the fact that most of the stakeholders included in the study, while involved in the topic to a greater or lesser extent, were not specialists in it. Therefore, most of the participants presumably had no overriding professional interests in palliative care. The response rate was better than in previous surveys on palliative care in Germany which involved health professionals [16]. This affirms the high level of interest in palliative care and end-of-life topics in society.

We found the basic acceptance of the aims of palliative care to be strong among the stakeholders, which might be helpful to the development of the charter process mentioned above [17]. However, the agreement of the health insurance funds' representatives to the majority of the palliative care issues presented was significantly lower than that of the palliative care organizations' representatives. This may result from a different awareness due to the usual absence of clinical experience on the part of health insurance funds' representatives, and also from the different aims and priorities due to the professional responsibilities and tasks in the health care system. The conclusion may be drawn that it is important to respect differing views, to learn from each other and to increase awareness of the specific palliative care approach among those who are not specialists.

Altogether there was consensus among the stakeholders about the meaningfulness of the measures for improvement presented. Compulsory training of health professionals in palliative care (items 1–3) met with the participants' greatest approval. This is not surprising as training deficits are among the major problems in palliative and end-of-life care in Germany despite advances in recent years, e.g. the implementation of the optional qualification in palliative medicine for physicians [6,9,18]. For example, at many universities palliative care is still not part of the curriculum and thus not regularly taught to medical students [9]. In contrast to its meaningfulness, the feasibility of compulsory training in palliative care (item 2) was given a far worse assessment. This is at first glance surprising, because it seems unclear why it should be so difficult to integrate palliative care training into the curricula for medical students. Maybe the already crowded curricula make it difficult to add further content. However, it has to be asked why this seems to be so difficult in Germany whereas it has been realized in other countries, e.g. France or Norway [6].

This brings up the key issue arising from the results: The feasibility of all measures was given a significantly worse assessment than their meaningfulness. Moreover, the meaningfulness of most measures (16 out of 18) was assessed positively whereas the feasibility of the measures in Germany was evaluated negatively (7 out of 18) or inconsistently (11 out of 18). Why does it appear to be so difficult to implement substantial changes in the German health system?

The German health system is based on local decision-making and the democratic legitimization of self-governing structures which are safeguards against unwanted government interference [19]. On the other hand, numerous players and lobby groups often represent different, conflicting interests; they may attempt to steer political developments in one particular direction or another in order to achieve advantages for their own pressure group. This complicates the political decision-making processes and may lead to substantial changes being implementable only with difficulty, since it is almost always the case that there are different lobby groups trying to promote different interests. As politicians are dependent on winning votes, lobby groups exercise great influence on voters and elections at local, state or federal level take place all the time in Germany and frequently lead to unstable political majorities, the result may be a certain degree of log-jam where major reforms in the field of health care are concerned.

For example, one serious weakness in the German system is the traditional fragmentation of care across the different health care sectors (e.g. inpatient and outpatient care, rehabilitation), which has been addressed by several recent reforms but has still not been satisfactorily overcome, resulting in serious difficulties in realizing substantial structural changes [13,15,19].

Another interesting issue is medical care for patients living in nursing homes. Up until now, medical care for patients in nursing homes has usually been carried out by the patients' individual family doctors without a fixed framework of cooperation with the nursing homes and without regular qualification in palliative care on the part of the doctors. The introduction of specialized nursing home physicians may be considered as an alternative. In our survey, this measure (item 14) was predominantly evaluated positively (80% "good") while the feasibility in Germany was evaluated negatively (22% "good").

It is worth mentioning that the medical associations agreed significantly less with the concept of nursing home physicians (65% "good") than did nursing organizations (100% "good") and patient organizations (96% "good"). The negative attitude on the part of the doctors' representatives found in our study is confirmed by recent official statements, e.g. by the German Medical Association: it is argued that nursing home physicians would not solve the main problem of the inappropriately low financing of the time-consuming medical care for nursing home residents [20]. However, with the recent nursing care insurance reform in 2008, the spectrum of possible improvement measures for medical care in nursing homes, which range from formal cooperation with family doctors and specialists to the appointment of nursing home physicians, was extended [21]. One may wonder whether these approaches will actually be made use of.

Also with the recent nursing care insurance reform, a legal right to six months' unpaid leave from work was introduced in Germany for the first time, if employees want to care for their relatives [21]. This measure (item 5) was favoured by nearly all of the participants in our study (92% "good"), but many of them considered the feasibility as being critical (25% "good"). Maybe they are afraid that barriers on the part of employers and employees (e.g. career disadvantages, and in the case of temporary work contracts) will impede realization in practice.

Case management performed by specialized case managers (item 15) tended to be assessed negatively, in particular by the medical associations' representatives. This is not surprising, as many physicians see it as part of their role to coordinate the health care of their patients [11]. However, shortcomings in coordination are among the most serious problems in Germany [19]. There is therefore no doubt that improvements in coordination and management are necessary; but the crucial question is: who is the most appropriate case manager and what form should the management take?

The literature shows that intensive home-based case management provided by registered nurse case managers may, in coordination with patients' existing sources of medical care, improve the realization of care of chronically severely ill patients in the last years of life [22]. Due to the traditional hierarchical structure and the significant cultural and educational barriers between physicians and nurses, it may be presumed that the acceptance of nurses as case managers on the part of the physicians will be difficult to achieve. However, nursing in Germany—as well as in many other countries—has been increasingly professionalized, specialized and academically qualified along with the achievement of new self-confidence [12].

In its recent report, the Advisory Council on the Assessment of Developments in the Health Care System (SVR) points out the need for role changes and new forms of cooperation among health professionals [23]. It is promising that the organization and management of specialized outpatient palliative care was explicitly implemented in the recent German health care reform in 2007 [10,11]. It is greatly to be hoped that appreciable improvements for the patients will result from the political efforts.

Limitations

There is one powerful lobby group not included in the study: the pharmaceutical industry. This plays an important role in the decision-making and policy-making processes for medical care, e.g. concerning the prices, distribution and availability of drugs within the system of the social health insurance that is responsible for approximately 90% of the citizens in Germany. It was discussed in our study group whether representatives from the pharmaceutical industry should also be surveyed, but we decided against it; we do not think that the industrial perspective should be part of the development of public health targets for palliative care.

Furthermore, it has to be considered that the results are not based on official statements from the institutions, organizations and associations, but reflect the personal attitudes and opinions of the representatives in combination with their professional background, experience and responsibility. We received many emails and telephone calls from participants who wondered if they should take part in the survey as they did not feel adequately qualified to evaluate the specific measures due to a lack of professional focus on palliative care. However, we encouraged them to participate, because it was important for us not to exclusively study experts' opinions. Rather, it was important for us to gain insight into the views of decision-makers and representatives from a wide range of socially and politically relevant groups. Also, we are convinced that a broad approach is needed for the further development of palliative care; that end-of-life questions affect everyone irrespective of their profession; that the professional background influences personal views and vice versa; and that personal experiences and attitudes play an important role in policy-making and decision-making.

Conclusion

If palliative care is to be improved in Germany, there are significant barriers to be overcome. Substantial improvement measures seem to be considered difficult to implement, possibly because of the traditional fragmentation of responsibilities and the partly conflicting interests of the different stakeholders. The recently initiated national charter process for the care of very severely ill and dying people is a promising approach to the further development of palliative care on a broad social, political and scientific basis. It may be helpful to the process that the basic understanding of the palliative care approach seems to be quite homogenous among many stakeholders. However, the awareness of the aims and content of palliative care should be increased on the part of those who are not specialists in palliative care.

Competing Interests

The authors declare that they have no competing interests.

Authors' Contributions

SLL and MB jointly recruited the participants, developed and tested the questionnaire, performed the statistical analyses and made contributions to the manuscript. SB made contributions to the conception and design of the study, the development of the questionnaire and the statistical analysis of data, and to the manuscript. NSCH conceived the study, supervised the recruitment, the development of the questionnaire and the analyses, interpreted the data and drafted the manuscript. All authors read and approved the final manuscript.

Acknowledgements

The study was funded by the German Research Foundation (DFG). In particular, we would like to thank all institutions and groups which participated in this study. Furthermore, we thank Prof. Marie-Luise Dierks for assisting in the development of the questionnaire.

References

1. Sepulveda C, Marlin A, Yoshida T, Ullrich A: Palliative care: the World Health Organization's global perspective. J Pain Symptom Manage 2002, 24:91–96.

2. Radbruch L, Nauck F, Ostgathe C, Elsner F, Bausewein C, Fuchs M, Lindena G, Neuwöhner K, Schulenberg D: What are the problems in palliative care? Results from a representative survey. Support Care Cancer 2003, 11:442–451.

3. National Institute for Clinical Excellence: Guidance on cancer services: improving supportive and palliative care for adults with cancer. Research evidence. London 2004.

4. Recommendation Rec(2003)24 of the Committee of Ministers to member states on the organisation of palliative care [http://www.coe.int/t/dg3/health/Source/Rec(2003)24_en.pdf].

5. World Health Organization: Palliative Care, the solid facts. Kopenhagen 2004.

6. Stand der Palliativmedizin und Hospizarbeit in Deutschland und im Vergleich zu anderen Staaten (Belgien, Frankreich, Großbritannien, Niederlande, Norwegen, Österreich, Polen, Schweden, Schweiz, Spanien) [http:/ / www.dgpalliativmedizin.de/ pdf/ Gutachten%20Jaspers-Schindler%20End fassung%20 50209.pdf].

7. Schindler T: Zur palliativmedizinischen Versorgungssituation in Deutschland. Bundesgesundheitsblatt 2006, 49:1077–1086.

8. Palliativmedizin & Hospizarbeit [http://www.dgpalliativmedizin.de]

9. Schneider N, Schwartz F: Hoher Entwicklungsbedarf und viele offene Fragen bei der Versorgung von Palliativpatienten. Med Klin 2006, 101:552–557.

10. Bekanntmachung eines Beschlusses des Gemeinsamen Bundesausschusses über die Erstfassung der Richtlinie zur Verordnung von spezialisierter ambulanter Palliativversorgung [http://www.g-ba.de/downloads/39-261-582/2007-12-20-SAPV-Neufassung_BAnz.pdf].

11. Schneider N: New specialist outpatient palliative care – a position paper (in German). Z Allg Med 2008, 84:232–235.

12. Ewers M, Schaeffer D: Dying in Germany – consequences of societal changes for palliative care and the health care system. J Public Health 2007, 15:457–465.

13. Diederichs C, Klotmann K, Schwartz F: The historical development of the German health care system and respective reform approaches. Bundesgesundheitsblatt 2008, 51:547–551.

14. Schwartz F, Kickbusch I, Wismar M: Institutionen, Systeme und Strukturen. In Das Public Health Buch. Gesundheit und Gesundheitswesen. Edited by: Schwartz F, Badura B, Busse R, Leidl R, Raspe H, Siegrist J, Walter U. München, Jena: Urban&Fischer; 2003:229–242.

15. Schneider N, Buser K, Amelung V: Improving palliative care in Germany: summative evaluation from experts' reports in Lower Saxony and Brandenburg. J Public Health 2006, 14:148–154.

16. Schneider N, Bisson S, Dierks M: Framework of palliative care in Germany and development of public health targets. Study design and methods [in German]. Bundesgesundheitsblatt 2008, 51:467–71.

17. Round table discusses charter for the care of very severely ill and dying people in Germany [http://www.bundesaerztekammer.de/page.asp?his=3.71.6895.68 96.6935&all=true]

18. Klaschik E: Palliativmedizin – Ganzheitliche Medizin mit hohem Entwicklungsbedarf. Schmerz 2001, 5:311.

19. Busse R, Riesberg A: Health Care Systems in Transition: Germany. World Health Organization Regional Office for Europe on behalf of the European Observatory on Health Systems and Policies. Kopenhagen; 2004.

20. Hibbeler B: Ärztliche Versorgung in Pflegeheimen. Von Kooperationen profitieren alle. Dtsch Arztebl 2007, 104:A3297–A3300.

21. Gesetz zur strukturellen Weiterentwicklung der Pflegeversicherung [http://www.bgblportal.de/BGBL/bgbl1f/bgbl108s0874.pdf].

22. Aiken L, Butner J, Lockhart C, Volk-Craft B, Hamilton G, Williams F: Outcome evaluation of a randomized trail of the PhoenixCare intervention: program of case management and coordinated care for the seriously chronically ill. J Palliat Med 2006, 9:111–126.

23. Cooperation and Responsibility – Prerequisites for Target-Oriented Health Care [http://www.svr-gesundheit.de/Gutachten/Gutacht07/KF2007-engl.pdf].

Knowledge and Communication Needs Assessment of Community Health Workers in a Developing Country: A Qualitative Study

Zaeem Haq and Assad Hafeez

ABSTRACT

Background

Primary health care is a set of health services that can meet the needs of the developing world. Community health workers act as a bridge between health system and community in providing this care. Appropriate knowledge and communication skills of the workers are key to their confidence and elementary for the success of the system. We conducted this study to document the perceptions of these workers on their knowledge and communication needs, image building through mass media and mechanisms for continued education.

Methods

Focus group discussions were held with health workers and their supervisors belonging to all the four provinces of the country and the Azad Jammu & Kashmir region. Self-response questionnaires were also used to obtain information on questions regarding their continued education.

Results

About four fifths of the respondents described their communication skills as moderately sufficient and wanted improvement. Knowledge on emerging health issues was insufficient and the respondents showed willingness to participate in their continued education. Media campaigns were successful in building the image of health workers as a credible source of health information.

Conclusion

A continued process should be ensured to provide opportunities to health workers to update their knowledge, sharpen communication skills and bring credibility to their persona as health educators.

Background

Primary Health Care (PHC) defined as "Essential health care made universally accessible to individuals and families in the community by means acceptable to them, through their full participation, and at a cost that community and country can afford" has been recommended as a set of health services that can meet the challenges of a changing world [1]. The World Health Organization (WHO) in its latest report has called for a revival of PHC [2].

An important component of the rejuvenated concept of PHC is community health workers, (CHWs) who act as a bridge between the health care delivery system and the community. Mary & Rosemary have described how CHWs enable health programmes to achieve three interconnected goals: building a relationship between the health care provider and laypersons in the community; improving appropriate health care utilization; and educating people to reduce health risks in their lives [3]. Highly challenging and innovative ideas such as serving 70% of a population of 190 million in Brazil, skin-to-skin care for newborns in India and improved perinatal care in Nepal have worked remarkably well through CHWs [4-6].

Appropriate knowledge and interpersonal communication expertise, in addition to basic clinical skills, supplies and supervision, are a key to the work of CHWs [7-9]. The CHW can empower the community to identify its needs and can assist in planning a strategy to achieve the desired results. In order to accomplish this successfully, CHWs should be culturally sensitive, with an ability to build a strong community rapport.

The 100 000 Lady Health Workers (LHWs) of Pakistan's Ministry of Health fit well into the definition of CHWs; their programme is considered as one of the successful large-scale community programmes [10]. Various evaluations have enumerated the successes of this programme, along with a few areas to ensure quality improvement [11,12]. Regarding quality improvement, a number of authors recommend devising strategies to improve health worker education and training, and suggest that preferences of primary care workers should be known and discussed at the policy level [13,14].

None of the evaluations from Pakistan have sought the workers' own perceptions regarding their knowledge and communication capacity, however. We therefore conducted this qualitative study in all provinces of Pakistan, to know the perceptions of health workers and supervisors on the communication capability of the LHW; adequacy of their knowledge; effectiveness of the image-building activities involving mass media; and mechanisms for continued education.

Methods

The Ministry of Health, Government of Pakistan, launched the National Programme for Family Planning and Primary Health Care of Pakistan, also called the Lady Health Workers Programme (LHWP) in 1994. Since then, the programme has deployed about 100 000 LHWs and more than 5000 Lady Health Supervisors (LHSs) in 135 districts of all the four provinces and regions of the country.

Providing appropriate and implementable health information to rural households has been the cornerstone of the programme's health promotion strategy. After recruitment, its workers undergo 15 months of preparation: three months of classroom training, followed by supervised fieldwork for one year. Refresher training sessions are also conducted from time to time.

There is a system of supervision meetings every month in which the 15 to 20 LHWs from the area share their progress and problems with the supervisor. The programme broadcasts issue-based communication campaigns on television and other mass media in which the LHW is positioned as an accessible and credible source of health information to the rural household.

We conducted a multi-stage, stratified, random sampling for this study. Under some of the donor-funded programmes, various initiatives for capacity building of the LHW are being carried out selectively throughout the country. To gauge the true programme situation, we selected only those districts where no donor-funded project was being implemented. These included two rural districts of Attock and Charsaddah from the provinces of Punjab and NWFP, and two urban districts of Karachi and Quetta from the provinces of Sind and Baluchistan. Muzaffarabad District was selected from the region of Azad Jammu & Kashmir (AJK), while tribal regions could not be considered because of the prevailing security situation in those areas. In each district, the sample comprised all LHWs and supervisors who were aged 20 to 50 years, based at their respective villages, married or unmarried, willing to participate in the study and having at least one year of work experience.

It was a cross-sectional study consisting of two components. Component 1 comprised focus group discussions (FGDs) with LHWs and their supervisors; in component 2, information from the same LHWs and their supervisors was obtained through a self-response questionnaire.

We developed guiding questions for the FGDs of component 1 and a self-reporting questionnaire for component 2. Careful attention was given to how the questions on "perceived barriers" would be asked during the FGD. We included appropriate examples to explain the questions uniformly across all the discussions. The questionnaires, originally developed in English, were translated to Urdu. The self-response questionnaire was back-translated as well, according to the recommendations [15]. The questionnaires were pilot-tested with groups in Rawalpindi District and appropriate changes made in the light of their feedback.

Prior appointments were made before travelling to the respective districts to conduct the discussions. Before formal discussion, the purpose of the research was explained to all the participants. The discussions held in Urdu and were tape-recorded after obtaining participants' permission. Notes were also taken, so that no discussion point was missed. Discussions were carried out with the help of guiding and probing questions. The self-reporting questionnaire was distributed among the participants at the end of each FGD. The questionnaire was first explained, after which the respondents filled in the required information and returned the questionnaires to the facilitator.

Inductive analysis [16,17] was performed on the transcripts and field notes through the following.

1. familiarization with the data, which included reading the notes, transcribing the FGDs and translating the Urdu transcript into English;
2. initial categorization by developing tables on themes emerging from the discussions. Notes taken during the discussions as well as the transcripts were consulted for developing these tables to ensure as much rigor as possible;
3. identifying patterns and connections within and between categories with the help of these tables;
4. entering data from the quantitative questionnaire into the Statistical Package for Social Sciences;
5. reaching a final interpretation by combining all the findings.

The answers of the majority were presented as the "Response," while comments that were significant but not shared by the majority were labelled as "Additional Comments." The study was undertaken between 8 March 2008 and 15 August 2008. Ethical approval for the study was obtained from the National Programme for Family Planning & Primary Health Care.

Results

Participants

A total of 105 participants, including 57 LHWs and 48 LHSs from five districts, participated in the research. They took part in FGDs and also filled out the questionnaire. The minimum number of participants in FGDs was seven (Karachi), while the maximum was 16 (Muzaffarabad). The mean age of the participant LHWs was 31 years (range 20–49). Among them, 88% were married, while 12% of the LHW were not married. The number of participating supervisors was 48, with a mean age of 30 years (range 23–50). Among the LHS, 92% were married, while 8% belonged to the unmarried category (Table 1).

Table 1. District-wise sociodemographic variables of the participants (n = 105)

District	LHW					LHS				
	No.	Age range, years	Mean age, years	Married	Unmarried	No.	Age range, years	Mean age, years	Married	Unmarried
Attock	12	24–40	30	9	3	9	27–32	30	9	0
Charsaddah	8	22–40	30	8	0	9	25–35	28	9	0
Karachi	8	28–49	34	8	0	7	23–39	32	6	1
Muzaffarabad	16	27–44	30	14	2	11	24–36	29	9	2
Quetta	13	20–45	32	11	2	12	21–50	29	11	1
Total	57	20–49	31	50	7	48	21–50	30	44	4

Communication Skills

Out of the five groups of LHWs, four believed they possessed moderately sufficient communication skills (Table 2). The group from Quetta, however, thought they possessed insufficient skills. The same proportion (four fifths) of the supervisors rated these skills as moderately sufficient, while one group (Karachi) called the communication skills of their LHW as sufficient.

Table 2. Responses to the questions regarding communication skills

Question		LHW	LHS
Are they sufficient?	Response	Moderate (4/5) Insufficient (1/5)	Sufficient (1/5) Moderate (4/5)
	Additional comments	There is room for improvement	LHWs with education <10 grades, not married, or those having low SES face more difficulty
How you deal with barriers perceived by individuals?	Response	By talking about child's future, using religious teachings, using help of influentials	Talking about child & family's future, using religion, using local leaders
	Additional comments	Using fear appeal and using help of LHS mentioned by some	Positive examples & using IEC materials, helping with own hand
What are your specific suggestions on the communications capacity building of LHW?	Response	Refresher training, role plays on common difficult scenarios, better IEC materials should be provided	Refresher training, role plays on common difficult scenarios, better IEC materials should be provided
	Additional comments	Adequacy and timeliness of the supply of IEC materials should be improved	Quality of basic training should also be improved

Communicating with males on family planning; establishing village health committees; convincing TB suspects to make use of diagnostic facilities; and talking about taboo subjects such as HIV/AIDS and other sexually transmitted diseases (STDs) were reported as health issues on which dialogue was difficult for the LHWs.

The respondents informed that nazr (evil eye), garam & thanda (hot & cold) food, male child preference, fear of stigma in TB and other diseases, and fatalism were the common barriers perceived by the community. Talking about the ways in which they addressed these barriers, the workers reported using better child health leading to better prospects for the family as an incentive for the people to make desired changes in their behaviour. They also reported using religious teachings where appropriate; using fear appeal; and seeking the help of influentials (teachers, counsellors, peers) where available. The main response of the majority of the supervisors was similar to that of the workers.

The respondents suggested refresher training sessions that include role plays on common difficult scenarios as a way to improve communication skills of the workers (Table 2). They proposed that appropriate information and skills to deal with people who were fixed on strong negative feelings, such as "we are poor, we can't do anything" or "a woman's only role is to serve the husband, kids and the family" or "the life or death of the mother or newborn is the will of God, in which the mortals cannot intervene" would be really helpful. The workers also suggested that information, education communication (IEC) materials should be provided to them that could be carried to the households and used for talking about specific health issues.

Level of Knowledge

Family planning (FP); maternal, newborn and child health (MNCH); nutrition; malaria; the Expanded Programme on Immunization (EPI) and common childhood diseases (Table 3) were reported as topics on which the workers had sufficient knowledge. Yet, they wanted more knowledge on some of these issues, e.g. MNCH, FP and communicable diseases such as TB. Emerging diseases such as Congo fever, avian influenza or dengue fever were reported as areas in which workers had insufficient knowledge. According to workers, their community asked questions on these emerging diseases to which they (workers) could not respond, as these topics were not a routine part of their curriculum or training.

Table 3. Respondents' views on adequacy of technical knowledge

Question	Response	
	LHW	LHS
What are the topics on which you have sufficient knowledge?	FP, MNCH, nutrition, malaria, EPI & common childhood diseases	FP, MNCH, nutrition, malaria, EPI, common childhood diseases & National Immunization Days
What are the topics on which you have insufficient knowledge?	Emerging diseases, medicinal issues, questions on repeated weighing and polio immunisation of babies are difficult topics	Emerging diseases, e.g. dengue fever, Congo fever, avian influenza, etc.

All the respondents thought some means of continuing education would help them improve their areas of weak knowledge. They liked the idea of receiving a regular publication from the programme. Out of the total, about 3% of the respondents showed interest in receiving official or administrative information and 40% were interested in reading clinical (tibbi maalooomat) information, while 57% wanted both types of information in equal amounts through such a publication.

The respondents were also asked whether a regular source of information, such as a periodical sent from the programme, would help them address the queries. A little over 94% thought such a regular source of information would help them respond to these questions (Figure 1).

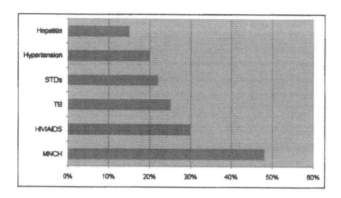

Figure 1. Six priority topics on which respondents sought further information (n = 105).

Media Campaigns

Respondents belonging to both categories in all the districts liked their representation in the mass media. Table 4 describes their views as well as suggestions on these media campaigns. They thought the media enhanced the credibility of health workers as the messages on television (TV), radio or newspapers were liked by their family and community members. According to them, people believed in the information provided by the LHW when a similar message was concurrently shown on TV. The supervisors thought the community respected the worker and acknowledged her services after having seen TV commercials (TVCs) about her roles and responsibilities. The respondents suggested some modifications to improve the mass media campaigns. In their view, using the genre of television drama, adding male characters to the media products and airing these campaigns on private cable TV channels in addition to the state-run terrestrial channels would help increase the effectiveness of these campaigns.

Table 4. Respondents views on media campaigns and its improvement

Question		LHW	LHS
What are your and your family/ community's views on media campaigns about the LHW programme?	Response	The campaigns please us/enhance our credibility.	The campaigns please us/enhance our credibility.
	Additional comments	People believe in our message when they have seen it on TV	Community relates to the worker because of these commercials
What changes should be made to improve these campaigns?	Response	Male characters should be added to TVCs. Drama format should also be tried. Multiple channels should be used	Role of LHS should be shown. Male characters should be added to TVCs.
	Additional comments	Community should also be shown.	Media should dispel that this programme is only about FP.

Discussion

To our knowledge, this is the first study that has explored how community-level health workers see their own performance. The workers and their supervisors acknowledged that there was room for improvement in their communication capacity. In spite of the attention given, knowledge on some of the areas that were part of the original curriculum remained weak, while at the same time the evolving public health situation in the country demanded addition of basic information on emerging health issues to the training system. The idea of a regular publication for continuing education of these workers therefore was highly appreciated.

Dealing with barriers perceived by the community requires communication skills in addition to updated knowledge. Interestingly, without guidelines the workers were using some of the recommended techniques, e.g. use of positive examples or fear appeal [18], but they wanted more capacity to deal with these barriers. Adding role plays in the training to deal with common difficult scenarios, as suggested by the health workers and their supervisors, could help the workers.

For the LHWs, talking to male members of their community about FP topped the list of "difficult to discuss" areas; participants from all over the country reported this difficulty. Given the conservative prevailing culture and the sensitivity of the topic, this difficulty is understandable, yet talking to males—who are the sole decision-makers in the patriarchal system of society—is vitally important. Adding male mobilizers to the health education arm can be one solution. Alternatively, the recently reported techniques [19] that employed the community worker to empower a woman to discuss with her husband vital issues such as child spacing and bring about change in the FP behaviour of the couple, should also be explored.

Bringing out a regular publication for the continued education of workers and their supervisors was a novel suggestion. Owing to the low level of literacy, the LHW in Pakistan is recruited with a minimal education of eight to 10 years of schooling. A three-month classroom training session is provided, which obviously is not enough to build her capacity to remember all the details on about 20 topics on which she is expected to talk. An additional burden is imposed by emerging health issues, which are not part of her curriculum but that become popular health topics when the fear of an epidemic arises.

Given the large number of LHWs, arranging frequent refresher training sessions to help them refresh their knowledge and gain information on new health issues also has many logistical and financial implications. In the light of this research, the programme has already initiated publishing a quarterly newsletter that contains both technical and administrative information and is mailed directly to all the 100 000 workers. Such innovations can be replicated by other CHW programmes.

The mass media messages disseminated by the programme from time to time brought recognition and credibility to the worker. After watching these campaigns on mass media, the community readily believed that these workers had been hired and trained by the Ministry of Health and would bring good advice and beneficial products. Low use of CHW programmes has been linked to poor community introduction of the programme [9]. Mass media campaigns have effectively addressed this issue with regard to the CHW programme in Pakistan. These campaigns, with suggested modifications, should be continued.

This research explored the views of health workers and their supervisors qualitatively as well as quantitatively, which is the strength of this study. However, as the respondents of both components were the same, their views can be called only suggestive, and not representative of the whole population. Similarly, how the community views the knowledge and communication capacity of these workers and their perceptions about the media campaigns conducted by the programme should also be explored, to develop a better understanding of the programme, its image and the performance.

In the context of resource constraints that many health systems face today, enhancing the role of the CHW has been highlighted as an alternative strategy by various experts [20,21]. According to WHO, the key factor in shortages of professional health workers in low-income countries can be addressed by "task-shifting," which is a delegation of tasks to the "lowest" category that can perform them successfully [9]. WHO has also recommended appropriate training and adequate and continuous support for these workers in order for them to perform optimally.

According to Nigel et al. [13], every country should strive to increase the number of health workers according to its priorities, but pragmatically many low-income countries initially focused on community and mid-level workers to address the high burden of disease in the primary care setting. They describe how Thailand improved its health system through this strategy during the 1970s–1990s and countries such as Brazil, Ethiopia, Ghana, India and Malawi have adopted a similar approach.

Conclusion

CHWs may seem elementary in high-resource settings, but they have a valuable role to play in developing countries. Some basic steps are required to facilitate them in improving their efficacy and effectiveness. A continued process should be ensured by primary health care programmes whereby opportunities are provided to community health workers to update their knowledge, sharpen communication skills and bring credibility to their persona as health educators.

Competing Interests

The authors declare that they have no competing interests.

Authors' Contributions

ZH and AH conceptualized this study; ZH carried out analysis of the data, conceptualized this paper and developed the first draft, while both authors jointly developed the final manuscript.

Acknowledgements

The authors are thankful to the federal, provincial and district staff of the National Program for FP & PHC, especially the LHWs and supervisors, for their help and participation, and to PAIMAN-USAID for supporting this study.

References

1. Alma Ata Declaration [http://www.who.int/hpr/NPH/docs/declaration_almaata.pdf].

2. World Health Organization: The World Health Report 2008: Primary Health Care Now More Than Ever. Geneva. 2008.

3. Mary AN, Rosemary S: State of evaluation: Community Health Workers. Public Health Nursing 2003, 20(4):260–270.

4. Bulletin of the World Health Organization [http://www.who.int/bulletin/volumes/86/4/08-030408.pdf].

5. Darmstadt GL, Kumar V, Yadav R, Singh V, Singh P, Mohanty S, Baqui AH, Bharti N, Gupta S, Misra RP, Awasthi S, Singh JV, Santosham M, the Saksham Study Group: Introduction of community-based skin-to-skin care in rural Uttar Pradesh, India. Journal of Perinatology 2006, 26:597–604.

6. Manandhar DS, Osrin D, Shrestha BP, Mesko N, Morrison J, Tumbahangphe KM, Tamang S, Thapa S, Shrestha D, Thapa B, Shrestha JR, Wade A, Borghi

J, Standing H, Manandhar M, Costello AM: Effect of a participatory intervention with women's groups on birth outcomes in Nepal: cluster-randomised controlled trial. The Lancet 2004, 364(9438):970–979.

7. Haq Z, Iqbal Z, Rahman A: Job stress among community health workers: a multi-method study from Pakistan. International Journal of Mental Health Systems 2008, 2:15.

8. Afsar H, Younus M: Recommendations to strengthen the role of lady health workers in the National Program for Family Planning and Primary Health Care in Pakistan: the health workers perspective. Journal of Ayub Medical College 2005, 17(1):48–53.

9. World Health Organization: Community Health Workers: What Do We Know About Them?. Evidence and Information for Policy, Department of Human Resources for Health, Geneva; 2007.

10. Haines A, Sanders D, Lehmann U, Rowe AK, Lawn JE, Jan S, Walker DG, Bhutta Z: Achieving child survival goals: potential contribution of community health workers. The Lancet 2007, 369:2121–31.

11. Ministry of Health Government of Pakistan: Internal Assessment of Lady Health Workers' Programme 2007. Islamabad: National Programme for Family Planning and Primary Health Care; 2008.

12. Oxford Policy Management: External Eevaluation of the National Programme for Family Planning and Primary Health Care: Summary of Final Report. Oxford. 2002.

13. Crisp N, Gawanas B, Imogen S: Training the health workforce: scaling up, saving lives. The Lancet 2008, 371:689–691.

14. Manongi RN, Marchant TC, Bygbjerg IC: Improving motivation among primary health care workers in Tanzania: a health worker perspective. Human Resources for Health 2006, 4:6.

15. Rahman A, Iqbal Z, Waheed W, Husain N: Translation and cultural adaptation of health questionnaires. Journal of Pakistan Medical Association 2003, 53:142–147.

16. Pope C, Zieblan S, Mays N: Qualitative research in health care: analyzing qualitative data. BMJ 2000, 320:114–116.

17. Taylor-Powell E, Renner M: Analyzing qualitative data. University of Wisconsin-Extension Program Development & Evaluation. [http://www.uwex.edu/ces/pdande]

18. Witte K: A meta analysis of fear appeals: implications for effective public health campaigns. Health Education & Behavior 2000, 27(5):591–615.

19. Rahman A, Malik A, Sikander S, Roberts C, Creed F: Cognitive behaviour therapy-based intervention by community health workers for mothers with depression and their infants in rural Pakistan: a cluster-randomised controlled trial. The Lancet 2008, 372:902–909.

20. Macinko J, de Souza MF, Guanais FC, Simoes CC: Going to scale with community-based primary care: an analysis of the family health program and infant mortality in Brazil, 1994–2004. Social Sciences & Medicine 2007, 65(10):2070–2080.

21. World Health Organization: The World Health Report 2006: Working Together For Health. Geneva. 2006.

Patient Experiences and Health System Responsiveness in South Africa

Karl Peltzer

ABSTRACT

Background

Patients' views are being given more and more importance in policy-making. Understanding populations' perceptions of quality of care is critical to developing measures to increase the utilization of primary health care services. Using the data from the South African World Health Survey (WHS), the current study aims to evaluate the degree of health care service responsiveness (both out-patient and in-patient) and comparing experiences of individuals who used public and private services in South Africa.

Methods

A population-based survey of 2352 participants (1116 men and 1236 women) was conducted in South Africa in 2003, the WHS—as part of a World Health Organization (WHO) project focused on health system performance assessment in member countries.

Results

Health care utilization was among those who attended in-patient care 72.2% attended a public and 24.3% a private facility, and of those who attended out-patient care 58.7% attended a public and 35.7% a private facility. Major components identified for out-patient care responsiveness in this survey were highly correlated with health care access, communication and autonomy, secondarily to dignity, confidentiality and quality of basic amenities, and thirdly to health problem solution. The degree of responsiveness with publicly provided care was in this study significantly lower than in private health care. Overall patient non-responsiveness for the public out-patient service was 16.8% and 3.2% for private care. Discrimination was also one of the principal reasons for non-responsiveness in all aspects of provided health care.

Conclusion

Health care access, communication, autonomy, and discriminatory experiences were identified as priority areas for actions to improve responsiveness of health care services in South Africa.

Background

The majority of South Africans depend on the public health sector for their health care needs [1]; the percentage of adults who reported that they have medical aid in the Demographic and Health Survey in 2003 was 14.2% [2]. Shisana et al. [3] found in a nationally representative survey that the majority (70%) indicated that they usually attended public health care services, while 23.3% attended private health care services, and a small proportion (0.1%) utilised traditional health practitioners. In many areas of South Africa, the Primary Health Care (PHC) facilities are the only available or easily accessible health service for local communities. As a result, PHC services, providers and facilities carry a large burden and responsibility for the provision of health care in South Africa. PHC is the basic mechanism for providing health care [4]. It was formally introduced in South Africa in April 1994 as the major principle for health care provision with the implementation of two policies, "Free health care for pregnant mothers and children under the age of six years" as well the "Universal Access to PHC for All South Africans" [4]. The Department of Health's strategic framework for 2002–2004 identifies improvements of quality of care as one of the four key challenges currently facing the health sector in South Africa [5]. Quality of care is concerned with the interface between provider and patients, between health services and community. A quality perspective changes the focus of health systems development from establishing structures to addressing what happens in the structures. Improving quality can, therefore, be regarded as a second phase of health care transformation in South Africa. The first phase was concerned with creating coherent health care structures and the second phase with ensuring quality of service delivery [5].

Variations in the perception of quality occur as a result of the heterogeneous nature of the definition of quality. Studies have pointed to variations in perception of quality by different socioeconomic groups as well as the environmental aspects such as the social, organisational and technological context of the service [6]. Van Vuuren and Botes [7] found among a culturally diverse population in an urban area in South Africa (greater Bloemfontein) that variables such as population group, age and employment status influence their attitudes towards professional health care. They further emphasise the importance of bringing these issues to the attention of health care policy makers. Peltzer [8] found in a community survey in rural South Africa a low acceptability of primary health care: 78% felt that the medical services are poor. Bediako, Nel and Hiemstra [9] found among hospital and out-patients in the North-West Province that more than half of the patients (56.8%) were not satisfied with the availability of medicines and other supplies. Approximately two thirds of patients (65.2%) did not know about the quality of telephone services rendered. There was a high level of dissatisfaction (63.1%) among patients regarding accessing doctors after hours. Most patients were satisfied with the general attitude of health workers (62.1%) but 21.2% were dissatisfied. De Jager and Du Plooy [10] found among in- and out-patients in a provincial hospital in Gauteng significant differences between in- and out-patients. Personal safety and cleanliness of facilities were regarded as the most important variables in the assurance and tangibility dimensions. The level of satisfaction was the highest for clear information and communication at an understandable level in the tangibility and assurance categories, respectively. The South African Department of Health [2] found that there was an increase in the percentage of adults who expressed dissatisfaction with all types of health services, except for traditional healers, comparing the Demographic and Health Survey (DHS) of 1998 and 2003. Generally, the results show that considerably more people are dissatisfied with the services rendered in hospitals, both public (23.3%) and private (11.6%). Even the levels of dissatisfaction with the services rendered by solo practitioners in the private sector (7.9%) seem to be on the increase during the period between the surveys. The major reasons for dissatisfaction with the public sector hospitals and community health centres were long waiting times (41.5% and 38.1% respectively), staff attitudes (22.8% and 25.9% respectively), non-availability of prescribed medication (15.8% and 17.7% respectively) and shortages of staff (doctors/pharmacists). Major reasons for dissatisfaction in the private hospital/clinic sector and private doctor were also long waiting times (26.7% and 7.4% respectively), staff attitude (18.0% and 7.1% respectively), and cost (15.2% and 24.8% respectively) [2]. Myburgh, Solanki, Smith and Lalloo [11] used a 1998 national population-based survey and found that both race and socio-economic status (SES) were significant predictors of levels of satisfaction with the services of the health care provider; White and high SES respondents were about 1.5 times more likely to report excellent service compared with African Black and low SES respondents, respectively.

Patients' views are being given more and more importance in policy-making. Understanding populations' perceptions of quality of care is critical to developing measures to increase the utilization of primary health care services.

A population-based survey was conducted in South Africa in 2003, the WHS—as part of a WHO project focused on health system performance assessment in member countries. Among the surveyed aspects was the evaluation of health care provision, based on the concept of "responsiveness" [12]. Using the data from the South African WHS, the current study aims to evaluate the degree of responsiveness with provided health care (both out-patient and in-patient), and comparing the experiences of individuals who used public and private services in South Africa.

Methods

Sample and Procedure

The country sample (n = 2352) was nationally representative and probabilistically selected using a multistage cluster design. All respondents were selected using a Kish table for selection within a household. The study included: (1) all individuals who had been hospitalized in the previous five years (stayed overnight in a hospital or other type of long-term care facility), and (2) among those who had not been hospitalized in the previous five years, all individuals who had used an out-patient health service in the past 12 months. Only the participants who had used an in-patient and/or out-patient health services were requested to complete the responsiveness questions. The number of responses obtained was a function of the overall response rate as well as the rate of service utilization in the previous 12 months for out-patient and 5 years for in-patient services. More detailed information about the World Health Survey is available on its website http://www.who. int/healthinfo/survey/en/index.html. To adjust for the population distribution as represented by the UN Statistical Division and for non-response, post-stratification corrections were made to the sampling weights.

Participants were interviewed face-to-face by lay people with at least a high school-level education; interviewers were trained in a week-long course using a standard manual and audiovisual aids as well as role-plays. Practice field interviews were reviewed by supervisors before actual data collection. All questionnaires were translated into major languages in South Africa and back-translated using a standard WHO protocol. The quality of translations was independently verified by bilingual experts before field implementation. Informed consent was obtained from all respondents and the study was cleared by ethics review committees.

Outcome Measures

The questionnaire included in this analysis used the health system responsiveness module. Responsiveness relates to patient's experiences with the health system, with a focus on the interpersonal aspects of the care, and differs from patient satisfaction, a construct that reflects people's expectations in addition to their experiences [13].

Questions covered the following aspects: traveling time to the health care provider (item wording: For your last visit [or hospital stay], how would you rate the traveling time to the health care provider [or hospital]?); waiting time before being attended to; being greeted and talked to respectfully; respect for intimacy during physical examination and care; clarity of explanations by the health care providers; availability of time to ask questions about the health problem or treatment; possibility of obtaining information on other types of treatment or tests; participation in decision-making on the health care or treatment; patient's freedom to speak privately with the health professionals; personal information kept confidential; freedom to choose the health care provider; inside the facility cleanliness including toilets; and available space in waiting and examination rooms. For participants who received in-patient care, two additional aspects were included: ease in receiving visits by family members (item wording: For your last hospital stay, how would you rate the ease of having family and friends visit you? and contact with the outside world) [13]. Response options for these responsiveness items were 1 = very good, 2 = good, 3 = moderate, 4 = bad, and 5 = very bad. These item responses were dichotomised into "1" and "2" = 1, and 3–5 = 0. In addition, there were three items for both out-patients and in-patients related to the health professional's skills (item wording for out-patients: In your opinion was the health care provider's skill adequate for your treatment? and item wording for in-patients: In your opinion, was the skill of the health care providers adequate for your treatment?), availability of medicines, and adequacy of equipment in the care; response options for these items were "yes" or "no." Cronbach alpha for the 16 items out-patients responsiveness scale was .89 for this sample and for the 18 items in-patient responsiveness scale .90 for this sample. Further, participants were asked whether they felt they had been treated worse by the health care providers (whether they felt discriminated) for on any of the following reasons: sex, age, lack of money, social class, ethnic group or skin color, type of illness, or nationality. Response options for these 7 items were "yes" or "no" [13].

Data Analysis

The first stage of this work included a descriptive analysis of the degree of responsiveness based on a set of variables that expressed the user's degree of experience, according to five response levels (1 = very good to 5 = very bad). For each of the items, the degree of responsiveness was estimated by the percentage of "good" or "very good" answers and the percentage of positive answers to three other dichotomous variables (yes or no), related to the health professional's skills, availability of medicines, and adequacy of equipment in the care. Percentages of patients' experiences were analyzed by "type of care" (out-patient or in-patient) and by public and private health care service utilization [13].

To explain total variation in the data set using a smaller number of factors, the second stage of the analysis used principal component analysis with varimax rotation.

Logistic regression was conducted for public versus private health care provider, and linear regression for the total responsiveness scale (16 responsiveness items for out-patients and 18 responsiveness items for in-patients), separately for public and private health care utilization. Demographic variables included sex, two age groups (18 to 39 and 40 years and above), two educational groups (up to primary school completed and above) [13], discriminatory experiences were summed up and converted into a binary variable with 1 = indicating any of the six discriminatory experiences and 0 = no such experiences.

For the statistical analysis the data were weighted according to the sampling design using Stata software version 10.0 (Stata Corporation, College Station, Texas, USA).

Results

The final sample included 2352 participants (1116 men and 1236 women) with a mean age of 37.6 years (SD = 14.3), range 18 to 97 years; the household response rate was 80% and the individual response rate was 90%. Average item missing rates were less than 1% (range 0–1.8%) for the responsiveness questions. Of the 2352 interviewees, 656 (28.3%) reported in-patient care in the five years prior to the survey. Of these, 72.2% attended a government public facility, 24.3% a privately operated health facility, 1.3% an NGO and 2.4% other. Among the participants, 449 (19.4%) had received outpatient care at least once in the year prior to the interview. Of these, 58.7% attended a government public facility, 35.7% a privately operated health facility, 0.9% an NGO and 4.7% other. Overall, 5.1% reported to have health insurance. The type of the last out-patient health care provider visited was 52.6% medical doctor, 36.9% nurse, 5.5% dentist, 2.5% physiotherapist and 2.4% other, none indicated traditional health practitioner.

According to the evaluation of out-patient care (see Table 1), "waiting time for care" showed the lowest degree of responsiveness (51.2%) among all the areas analyzed. While, the aspects related to health professionals' skills (92.3%), adequate equipment (91.6%), adequate availability of medicines received (87.1%), and respect for intimacy during physical examination (77.9%) had the highest responsiveness scores. The percentage of users who gave "good" or "very good" ratings was lower among users of the public health care system for all aspects studied, both for out-patient and inpatient care. The percentage of users who rated their experiences with "bad" or "very bad" was 16.8% for public and 3.2% for private out-patient care (see Table 1).

Table 1. Last visit responsiveness in percent by public or private out-patient care (n = 424, excluding NGO and other: n = 25)

Out-patient care	Total (n = 424)	Public (n = 367)	Private (n = 157)	OR (95% CI)
Age (Mean)	37.6	41.0	36.6	Public = 1.00 Private = 1.02 (1.01–1.04)***
Female patient	61.6	69.1	49.2	Public = 1.00 Private = 2.31 (1.49–3.59)***
Education (Mean, range 1–7)	3.8	3.3	4.7	Public = 1.00 Private = 0.56 (0.48–0.66)***
Seen by medical doctor	53.5	36.6	81.7	Public = 1.00 Private = 8.13 (8.08–8.22)***
Time to get to facility in minutes (Mean)	29.3	33.5	21.0	Public = 1.00 Private = 1.02 (1.00–1.04)*
Provider fees (Mean in Rand)	37.5	16.1	150.1	Public = 1.00 Private = 0.98 (0.97–0.99)**
Medicines costs (Mean in Rand)	16.2	6.5	38.1	Public = 1.00 Private = 0.97 (0.94–1.00)*
Patient satisfaction				
1. Traveling time to the health care provider	64.6	54.3	81.0	Public = 1.00 Private = 0.38 (0.17–0.46)***
2. Waiting time	51.2	36.2	75.4	Public = 1.00 Private = 0.19 (0.12–0.30)***
3. Being greeted and talked to respectfully	69.2	59.7	84.7	Public = 1.00 Private = 0.81 (0.26–2.48)
4. Respect for intimacy during physical examination	73.6	65.2	87.2	Public = 1.00 Private = 0.60 (0.18–1.99)
5. Clarity of explanations	73.2	63.6	88.9	Public = 1.00 Private = 0.22 (0.12–0.39)***
6. Time for questions	64.5	52.7	83.7	Public = 1.00 Private = 0.22 (0.13–0.37)***
7. Possibility of obtaining information on other types of treatment	61.0	49.2	80.2	Public = 1.00 Private = 0.24 (0.15–0.39)***
8. Participation in treatment decision-making	58.0	48.5	73.5	Public = 1.00 Private = 0.34 (0.21–0.54)***
9. Privacy with health professionals	68.7	59.5	83.6	Public = 1.00 Private = 0.29 (0.17–0.49)***
10. Confidentiality of personal information	78.4	71.1	99.7	Public = 1.00 Private = 0.28 (0.15–0.54)***
11. Freedom to choose health care provider	56.9	43.2	80.8	Public = 1.00 Private = 0.17 (0.11–0.28)***

Table 1. *(Continued)*

12. Cleanliness inside the health facility	72.7	62.6	89.1	Public = 1.00 Private = 0.21 (0.11–0.30)***
13. Available space in waiting and examination rooms	63.3	51.1	83.1	Public = 1.00 Private = 0.21 (0.13–0.36)***
14. Satisfactory health care provider skills	92.3	89.1	97.7	Public = 1.00 Private = 0.32 (0.13–0.81)*
15. Adequate equipment	91.6	89.5	94.9	Public = 1.00 Private = 0.52 (0.24–1.13)
16. Adequate availability of medicines	87.1	84.6	91.0	Public = 1.00 Private = 0.52 (0.31–1.00)
Total mean responsiveness (range 0–16)	11.4	9.8	13.6	Public = 1.00 Private = 0.78 (0.73–0.83)***

***$p < .001$; **$p < .01$; *$p < .05$

Among all the aspects of in-patient care responsiveness (Table 2), the lowest percentage of responsiveness was attributed to "freedom to choose the health care provider" (55.7%), while areas related to adequate health care provider skills, equipment, and availability of medicines had the highest health system responsiveness.

A significant proportion of out-patient care users experienced discrimination for the following reasons: 11.9% reported feeling they had been treated worse than others because of lack of money and 9.3% because of their social class. Among users of the public services these figures were 15.9% and 13.3%, respectively. Of all users, 6.4% reported they had been treated worse because of their skin colour. Interviewees who had been hospitalized in the previous five years reported lower discrimination rates than out-patients, with "lack of money," race and "social class" as major factors. Users of in-patient and out-patient public health care reported significantly higher discrimination rates than private health care patients (see Table 3 and 4).

Principal component analysis found for out-patient care responsiveness three main factors explaining 58.5% of the variance and for in-patient care responsiveness four factors explaining 62.9% of the variance. The three factors for out-patient care responsiveness included 1) six items with two each on time, communication and autonomy (explaining 38.6% of the variance), 2) six items with each two items on dignity, confidentiality and quality of basic amenities (explaining 12.5% of the variance), and 3) three items on health problem solution (explaining 7.4% of the variance). The four factors for the in-patient care satisfaction included 1) six items with two each on communication, autonomy and confidentiality (explaining 37.1% of the variance), 2) four items with two each on time and dignity (explaining 11.7% of the variance), 3) three items on health problem solution (explaining 8.0% of the variance), and 4) four items with two each on quality of basic amenities and access to family and community support (explaining 6.2% of the variance). The overall responsiveness score was for out-patients 67% and for in-patients 68% (see Table 5).

Table 2. Last visit responsiveness in percent by public and private in-patient care (n = 633, excluding NGO and other: n = 23)

In-patient care	Total (n = 633)	Public (n = 472)	Private (n = 161)	OR (95% CI)
Age (Mean)	38.9	38.7	39.4	Public = 1.00 Private = 0.99 (0.98–1.05)
Female patient	60.4	63.3	51.7	Public = 1.00 Private = 1.61 (1.49–3.59)[a]
Education (Mean, range 1–7)	3.9	3.5	5.1	Public = 1.00 Private = 0.40 (0.34–0.48)[***]
Time to get to facility in minutes	37	42	26	Public = 1.00 Private = 1.02 (1.00–1.03)[a]
Provider fees (in Rand)	246	44	907	Public = 1.00 Private = 0.99 (0.99–1.00)[***]
Medicines costs (in Rand)	78	9	299	Public = 1.00 Private = 0.99 (0.98–0.99)[***]
Tests	32	7	115	Public = 1.00 Private = 0.99 (0.99–1.00)[***]
Transport	37	41	24	Public = 1.00 Private = 1.00 (0.99–1.01)
Number of people slept in same room (Mean)	5.8	6.6	3.5	Public = 1.00 Private = 1.22 (1.06–1.40)[**]
Patient satisfaction				
1. Travelling time to the health care provider	67.8	61.2	87.5	Public = 1.00 Private = 0.23 (0.14–0.38)[***]
2. Waiting time	63.3	55.2	87.1	Public = 1.00 Private = 0.18 (0.11–0.31)[***]
3. Being greeted and talked to respectfully	70.6	63.3	92.0	Public = 1.00 Private = 0.15 (0.08–0.28)[***]
4. Respect for intimacy during physical examination	77.9	72.1	94.5	Public = 1.00 Private = 0.14 (0.07–0.30)[***]
5. Clarity of explanations	71.0	63.9	91.7	Public = 1.00 Private = 0.16 (0.08–0.31)[***]
6. Time for questions	63.0	55.0	86.4	Public = 1.00 Private = 0.19 (0.11–0.33)[***]
7. Possibility of obtaining information on other types of treatment	61.2	53.4	84.6	Public = 1.00 Private = 0.21 (0.13–0.35)[***]
8. Participation in treatment decision-making	60.6	50.4	90.6	Public = 1.00 Private = 0.11 (0.06–0.19)[***]
9. Privacy with health professionals	69.5	62.3	91.0	Public = 1.00 Private = 0.16 (0.09–0.30)[***]

Table 2. *(Continued)*

10. Confidentiality of personal information	76.9	71.3	93.0	Public = 1.00 Private = 0.19 (0.10–0.36)***
11. Freedom to choose health care provider	55.7	45.8	84.4	Public = 1.00 Private = 0.11 (0.09–0.26)***
12. Cleanliness inside the health facility	72.6	65.8	92.6	Public = 1.00 Private = 0.15 (0.08–0.30)***
13. Available space in waiting and examination rooms	67.8	60.0	90.9	Public = 1.00 Private = 0.12 (0.07–0.22)***
14. Satisfactory health care provider skills	93.7	92.3	97.8	Public = 1.00 Private = 0.29 (0.09–0.88)***
15. Adequate equipment	91.8	89.9	97.2	Public = 1.00 Private = 0.26 (0.10–0.72)**
16. Adequate availability of medicines	89.7	87.8	95.3	Public = 1.00 Private = 0.44 (0.23–0.86)*
17. Ease of receiving visitors	73.9	67.1	94.1	Public = 1.00 Private = 0.13 (0.06–0.26)***
18. Ease of staying in contact with outside world	62.4	52.9	90.3	Public = 1.00 Private = 0.12 (0.08–0.88)***
Total mean responsiveness (range 0–18)	12.9	11.7	16.3	Public = 1.00 Private = 0.70 (0.64–0.77)***

***$p < .001$; **$p < .01$; *$p < .05$

Table 3. Percentage of patients who experienced some type of discrimination by public and private out-patient care

Reason for discrimination	Total	Public	Private	OR (95% CI)
Out-patient care				
Sex	3.3	4.9	0.7	
Age	5.9	8.8	1.3	
Lack of money	11.9	15.9	5.3	
Social class	9.3	13.3	3.0	
Race	6.4	7.8	4.2	
Type of illness	6.3	9.0	2.0	
Nationality	4.3	5.7	2.1	
Total (Mean, range 0–7)	0.44	0.57	0.21	Private = 1.00 Public = 1.49 (1.09–2.03)*

The results of the multivariate analysis of the joint influence of sex, age group, formal education, public versus private health care, and discrimination experience on health care responsiveness are presented in Table 6. Only private health service and lower discrimination experienced were consistently associated with the total patient satisfaction score.

Table 4. Percentage of patients who experienced some type of discrimination by public and private in-patient care

Reason for discrimination	Total	Public	Private	OR (95% CI)
Sex	2.2	2.8	0.3	
Age	3.1	3.9	0.3	
Lack of money	8.2	10.4	1.5	
Social class	6.8	8.6	1.7	
Race	7.0	8.3	3.2	
Type of illness	4.4	5.6	0.6	
Nationality	2.1	2.3	1.6	
Total (Mean, range 0–7)	0.35	0.43	0.11	Private = 1.00 Public = 1.80 (1.17–2.76)**

Table 5. Principal component analysis with varimax rotation of health care responsiveness by out-patient and in-patient care (only items loading .40 or more are recorded)

	Frequency (%) Out-patient = OP In-patient = IP	Outpatient experience			In-patient experience			
		Factor 1 Time/ Communication /Autonomy	Factor 2 Dignity/ Confidentiality/ Basic amenities	Factor 3 Health problem solution	Factor 1 Communication/ Autonomy/ Confidentiality	Factor 2 Time/dignity	Factor 3 Health problem solution	Factor 4 Basic amenities/ supports
Travelling time to the health care provider	Time OP = 57.9	.45				.49		
Waiting time	IP = 65.6	.49				.76		
Being greeted and talked to respectfully	Dignity		.55			.60		
Respect for intimacy during physical examination	OP = 71.4 IP = 74.3		.72			.49		
Clarity of explanations	Communication	.73			.73			
Time for questions	OP = 68.9 IP = 67.8	.80			.77			
Possibility of obtaining information on other types of treatment	Autonomy OP = 59.5 IP = 60.9	.79			.76			
Participation in treatment decision-making		.80			.70			
Privacy with health professionals	Confidentiality		.53		.70			
Confidentiality of personal information	OP = 72.6 IP = 73.2		.56		.53			
Cleanliness inside the health facility	Quality of basic amenities		.80	.55				.46
Available space in waiting and examination rooms	OP = 68.0 IP = 70.2		.72					.48
Satisfactory health care provider skills	Health problem solution OP = 90.3 IP = 91.7			.78			.85	

Table 5. (Continued)

Adequate equipment			.88				.89	
Adequate availability of medicines			.74				.76	
Ease of receiving visitors	Support IF = 66.2							.67
Ease of staying in contact with outside world								.69
Summary	OP = 67 IP = 68							
Variance (%)		38.6	12.5	7.4	37.1	11.7	8.0	6.2

Table 6. Multivariate linear regression of demographic and health variables on total patient satisfaction [Dependent Variable: Ambulatory or in-patient responsiveness]

	Out-patients		In-patients	
	Total patient satisfaction		Total patient satisfaction	
	Coef. (CI 95%)	P	Coef. (CI 95%)	P
Sex Male Female	-.21 (-1.02–0.61)	0.61	-.43 (-1.17–.30)	0.24
Age 18–39 40 and more	.04 (-.83 – 0.91)	0.93	.14 (-.59–.86)	0.71
Education 1–3 4–7	.23 (-.69–1.14)	0.62	.77 (-.08–1.62)	0.07
Form of payment Public Private	-3.20 (-4.07–-2.33)	.000	-3.95 (-4.73–-3.18)	0.000
Discrimination Yes No	-3.90 (-5.03–-2.78)	.000	-3.99 (-5.02–-2.95)	0.000

Discussion

The study conducted among a nationally representative sample in South Africa found that of those who attended in-patient care 72.2% attended a public and 24.3% a private facility, and of those who attended out-patient care 58.7% attended a public and 35.7% a private facility, none indicated traditional health practitioner. Similarly, Shisana et al. [3] found in a nationally representative study that the majority (70%) indicated that they usually attended public health care services, 23.3% attended private health care services, and 0.1% utilised traditional health practitioners.

The international comparison of health care responsiveness using the same measures and analysis found that overall South Africa (67% for out-patients and 68% for in-patients) had much lower responsiveness than Brazil (80% for out-patients and 76% for in-patients) and Israel and 14 European countries (81% and higher) for both out-patient and in-patient care. Looking at different components of responsiveness, South Africa scored particularly low on waiting time (58% for out-patients and 66% for in-patients) and autonomy (60% for out-patients and 61% for in-patients) care compared to Brazil (65% and 69% for out-patient and in-patient experiences respectively for waiting time and 70% and 66% respectively for autonomy), Israel (69% and 77% for waiting time and 80% and 79% for autonomy) and European countries (72% and 81% for waiting time and 83% and 72% for autonomy. The relative rankings of the domains among out-patients were similar in South Africa and Brazil, with the three highest rankings on confidentiality, dignity and communication, and the three lowest on waiting time, autonomy and quality of basic amenities. Regarding relative rankings on the domains among out-patients rankings were similar in South Africa and Brazil, with quality of basic amenities ranking third, after dignity and confidence. Rankings for European countries were similar, with the exception that quality of basic amenities was ranked higher than in middle income countries (Brazil and South Africa) (see Table 7) [13-15].

Table 7. Health care responsiveness (percentage of respondents who responded either "good" or "very good") comparisons across countries [13-15]

	Out-patient experiences				In-patient experiences			
	South Africa	Brazil	Israel	European countries*	South Africa	Brazil	Israel	European countries*
Time	58	65	69	72	66	69	77	81
Dignity	71	93	92	90	74	90	90	89
Communication	69	81	87	87	67	76	87	82
Autonomy	60	70	80	83	61	66	79	72
Confidentiality	74	90	88	89	73	80	83	82
Quality of basic amenities	68	80	90	91	70	80	60	87
Support					68	70	91	92
Summary	67	80	83	87	68	76	81	83

*European countries included were: Austria, Belgium, Denmark, Finland, France, Germany, Greece, Ireland, Italy, Luxemburg, Netherlands, Portugal, Sweden, and United Kingdom

Principal component analysis found in this study that responsiveness in out-patients included as major factors waiting time/communication/autonomy followed by dignity/confidentiality/basic amenities, and for in-patients communication/

autonomy/confidentiality, waiting time/dignity and lastly quality of basic amenities/support. Similarly, Valentine, Darby and Bonsel [16] found from general population surveys of "health systems responsiveness" in 41 countries that most respondents selected prompt attention as the most important domain. Dignity was selected second, followed by communication. Access to social support networks was identified as the least important domain. The factor solutions from this study did not confirm the domain structure of 7 domains of previous studies [17]. This study underlines different clustering patterns of responsiveness for outpatients and in-patients in South Africa, in the Brazilian WHYS study [13] and in a study in Taiwan that found five factors (respect, access, confidentiality, basic amenities, and social support) [18]. For example, "autonomy" was in this and the Taiwanese study [18] not conceptualized as a unique domain. Further, studies are needed to identify the structure of health systems responsiveness domains in developing countries.

Regarding health care provider skills among out-patients in this study similar results were found between this study and the Brazilian WHS: 92.3% and 92.9% respectively, adequate equipment 91.3% and 91.6% respectively, and adequate availability of medicines 80.8% and 87.1%, and also among inpatients 91.3% and 93.7% for satisfactory health care provider skills, 92.3% and 91.8% for adequate equipment, and 92.9 and 89.7 for availability of medicine [13].

Major components identified for out-patient care responsiveness in this survey were highly correlated with health care access, communication and autonomy, secondarily to dignity, confidentiality and quality of basic amenities, and thirdly to health problem solution. Thus, from the perspective of health service users in South Africa, health care responsiveness was primarily related to health care access, communication and autonomy. Each of the three components got the lowest responsiveness ratings (59%–77%) compared to the other components (dignity, confidentiality, basic amenities and health problem solution) (76%–84%).

The degree of responsiveness with publicly provided care was in this study significantly lower than in private health care; a finding which was also found in local studies [2,8] and in the Brazilian WHS [13]. Overall lack of responsiveness for the public out-patient service was 16.8% in this study, which is lower to the DHS survey (23.3%) measuring patient dissatisfaction [2]. Both studies were conducted in the same year, 2003. Possible explanations for the above differences may lie in the better quality of private services or that expectations are already high among the population, both for users and nonusers, and because the different measures used in terms of getting a lower score with the responsiveness measure as compared to a dissatisfaction measure. In this study 15.4% of public out-patients were dissatisfied with the availability of medicines, which seem lower than in some local studies, 56.8% [9]. In multivariate regression analysis sex, age and educational level were not found to be associated with health care responsiveness unlike in some other studies [7,11,13,17].

Another problem identified in this study and also confirmed in the Brazilian WHS [13] was the high percentage of individuals who felt discrimination, regardless of public or private health care. Discrimination was also one of the principal reasons for dissatisfaction in all aspects of provided health care. The principal sources of discrimination identified by respondents were lack of money, social class and race. Gueveiva et al. [13] also found among the Brazilian WHS lack of money and social class as major factors of health care discrimination. It is important to note that the percentages of individuals who felt they had been treated worse than others on grounds of social exclusion were consistently higher among users of the public health care system, a practice that runs counter to the Bhato Pele (People first) guiding principles of the South African health care system. According to a qualitative study by Mashego and Peltzer [19] discrimination was also identified among primary public care users.

Unlike in some other studies [7], this study did not find significant associations between socio-demographic variables (age, sex and formal education) and patient satisfaction.

Conclusion

Health care access, communication, autonomy, and discriminatory experiences were identified as priority areas for actions to improve responsiveness and patient satisfaction in South Africa. Implications for policymaking include that the result from the survey can be used to prioritize efforts when resources are limited. The data seem to provide a clear message to prioritize reforms that improve prompt attention, but not at the expense of patient dignity and communication, which may damage the acceptability of health services to users, and result in barriers to access.

Study Limitations

In this survey all respondents who had been hospitalized in the five years prior to the survey were asked about their most recent hospitalization, and all others were asked about their most recent ambulatory visit in the previous year. The result is that the set of respondents who answered the questions about ambulatory care is small and not representative of the general population, but rather of the population that had not been hospitalized during the previous five years. The cross-sectional study design did not permit an investigation of the cause-effect relationship between responsiveness and independent variables. Recall bias of study participants cannot be excluded, especially on the 5 year recall period for hospital admission.

Competing Interests

The author declares that they have no competing interests.

Author's Contributions

KP designed the study, conducted the secondary analysis, and drafted and corrected the paper.

Acknowledgements

The paper uses data from the WHO World Health Surveys Study.

References

1. Viljoen R, Heunis C, van Rensburg EJ, van Rensburg D, Engelbrecht M, Fourie A, Steyn F, Matebesi Z: National primary health care facilities survey. Bloemfontein: Centre for Health Systems Research & Development, University of the Free State; 2000.

2. Department of Health: South African Demographic and Health Survey 2003. Pretoria: Department of Health; 2007.

3. Shisana O, Rehle T, Simbayi LC, Parker W, Zuma K, Bhana A, Connolly C, Jooste J, Pillay V: South African national HIV prevalence, HIV incidence, behaviour and communication survey. Cape Town: HSRC Press; 2005.

4. Department of Health: Primary health care progress report: health monitoring and evaluation. Pretoria: Department of Health; 2000.

5. Department of Public Services and Administration: Transforming Public Service, Delivery White Paper (Batho Pele White Paper), (Gazette 18340 Notice 1459). Pretoria, South Africa: Department of Public Services and Administration; 1997.

6. Goldstein S, Price M: Utilisation of primary curative services in Diepkloof, Soweto. S Afr Med J 1995, 85(6):505–8.

7. Van Vuuren SJEJ, Botes LJS: Attitudes towards health care in greater Bloemfontein. Curationis 1994, 17:2–10.

8. Peltzer K: Community perceptions of biomedical health care in a rural area in the Limpopo Province South Africa. Health SA Gesondheid 2000, 5(1):55–63.

9. Bediako MA, Nel M, Hiemstra LA: Patients' satisfaction with government health care and services in the Taung district, North West Province. Curationis 2006, 29(2):12–5.

10. De Jager J, Du Plooy T: Service quality assurance and tangibility for public health care in South Africa. Acta Commercii 2007, 7:96–17.

11. Myburgh NG, Solanki GC, Smith MJ, Lalloo R: Patient satisfaction with health care providers in South Africa: the influences of race and socioeconomic status. Int J Qual Health Care 2005, 17(6):473–7.

12. Üstun TB, Chatterji S, Villanueva M, Celik LBC, Sadana R, Valentine N, Oritz J, Tandon A, Salomon J, Cao Y, Wan Jun X, Özaltin E, Mathers C, Murray CJL: Multi-country Survey Study on Health and Responsiveness 2000–2001. Geneva: World Health Organization; 2001. GPE Discussion Paper 37

13. De Souza WV, Luna CF, De Souza-Júnior PRB, Szwarcwald CL: Health care users' satisfaction in Brazil, 2003. Cad Saúde Pública 2005, 21(Sup):S109–118.

14. Valentine NB, Ortiz JP, Tandon A, Kawabata K, Evans DB, Christopher JL, Murray CJL: Patient experiences with health services: population surveys from 16 OECD countries. In Health systems performance assessment: debates, methods and empiricism. Edited by: Murray CJL, Evans DB. Geneva: WHO; 2003:643–652.

15. Goldwag R, Rosen B: Responsiveness of the health care system: findings from the Israeli component of the World Health Survey. Jerusalem: Myers-JDC-Broodale Institute; 2007.

16. Valentine N, Darby C, Bonsel GJ: Which aspects of non-clinical quality of care are most important? Results from WHO's general population surveys of "health systems responsiveness" in 41 countries. Soc Sci Med 2008, 66(9):1939–50.

17. Valentine NB, Bonsel GJ, Murray CJ: Measuring quality of health care from the user's perspective in 41 countries: psychometric properties of WHO's questions on health systems responsiveness. Qual Life Res 2007, 16(7):1107–25.

18. Hsu CC, Chen L, Hu YW, Yip W, Shu CC: The dimensions of responsiveness of a health system: a Taiwanese perspective. BMC Public Health 2006, 17(6):72.

19. Mashego T-AB, Peltzer K: Community perception of quality of (primary) health care services in a rural area of Limpopo Province, South Africa: a qualitative study. Curationis 2005, 28(2):13–21.

Malnutrition Prevalence and Precision in Nutritional Care Differed in Relation to Hospital Volume – A Cross-Sectional Survey

Albert Westergren, Christine Wann-Hansson,
Elisabet Bergh Börgdal, Jeanette Sjölander,
Rosmarie Strömblad, Rosemarie Klevsgård,
Carolina Axelsson, Christina Lindholm and Kerstin Ulander

ABSTRACT

Background

To explore the point prevalence of the risk of malnutrition and the targeting of nutritional interventions in relation to undernutrition risk and hospital volume.

Methods

A cross-sectional survey performed in nine hospitals including 2 170 (82.8%) patients that agreed to participate. The hospitals were divided into large, middle, and small sized hospitals. Undernutrition risk and overweight (including obesity) were assessed.

Results

The point prevalence of moderate/high undernutrition risk was 34%, 26% and 22% in large, middle and small sized hospitals respectively. The corresponding figures for overweight were 38%, 43% and 42%. The targeting of nutritional interventions in relation to moderate/high undernutrition risk was, depending on hospital size, that 7–17% got Protein- and Energy Enriched food (PE-food), 43–54% got oral supplements, 8–22% got artificial nutrition, and 14–20% received eating assistance. Eating assistance was provided to a greater extent and artificial feeding to a lesser extent in small compared to in middle and large sized hospitals.

Conclusion

The prevalence of malnutrition risk and the precision in provision of nutritional care differed significantly depending on hospital volume, i.e. case mix. It can be recommended that greater efforts should be taken to increase the use of PE-food and oral supplements for patients with eating problems in order to prevent or treat undernutrition. A great effort needs to be taken in order to also decrease the occurrence of overweight.

Background

Nutritional screening is important in identifying persons who require treatment, as malnutrition is an under-recognised and under-treated condition [1-3]. Nutritional screening, assessment and treatment is emphasised in a resolution from the Council of Europe [4] and Swedish hospitals have been found to have difficulties in living up to the recommendations [5]. However, knowledge about the correspondence between the findings from nutritional screening and the actual provision of treatment is sparse. From a general perspective, knowledge about the prevalence of risk of malnutrition, i.e. undernutrition risk and overweight (including obesity), is important as it describes the magnitude of these problems and has implications for allocating health care resources for helping patients to remain or become well nourished.

Several factors influence the prevalence of risk of malnutrition. Among these are type of patients, i.e. case mix, within the different hospital settings and also the methods used to measure undernutrition or overweight. This can make it difficult to compare between different hospitals. To give some examples, it was found that 27% of the patients in middle and small sized hospitals (< 500 beds) were at risk of undernutrition (which means at least two of: unintentional weight loss, low Body Mass Index (BMI) and eating difficulties) [6], while in a teaching hospital it was 27% (based on anthropometry) or 46% (based on Subjective Global Assessment (SGA)) that were at undernutrition risk [7]. In a national screening in the Netherlands, 26% of the patients were found to be undernourished (weight loss) including all types of hospitals [8]. In a British study, 26% were at medium or high risk of undernutrition in smaller hospitals (< 1000 beds) and 38% in larger hospitals (> 1000 beds, Malnutrition Universal Screening Tool) [9]. The opposite was found in a nationwide German hospital study were 27% were malnourished according to SGA, and more were undernourished in smaller (37%) than in larger hospitals (20%) [10]. Oncological, gastrointestinal and lung diseases are associated with the highest prevalence of undernutrition [8].

It is not only undernutrition risk that is a problem in hospitals, but also overweight. The prevalence of overweight has also been presented previously, and in the British survey 52% were found to be overweight (BMI > 25 kg/m^2) [9]. In another study [6], the prevalence of overweight in middle and small sized hospitals was found to be 39% (BMI =/> 25 if </= 69 yrs, BMI =/> 27 if =/> 70 yrs). The studies referred to indicate that there are differences in the prevalence of risk of malnutrition in relation to type of hospital and the case mix therein. In addition, the prevalence found in different studies varies due to differences in definitions and the use of different screening tools [11]. Thus, it seems worthwhile to explore the prevalence of undernutrition risk as well as overweight using the same criteria (one screening tool), in relation to hospital volume.

Besides exploring the prevalence of undernutrition risk and overweight, it is important to gain knowledge about how well nutritional interventions are targeted to those at undernutrition risk. This is especially interesting with regard to patients with energy problems (who eat little, stop eating due to tiredness rather than to having satisfied their hunger, eat slowly and/or have poor appetite), especially as these types of problems have been found to have a negative impact on nutritional status and are related to the provision of protein- and energy enriched food and oral supplements [12,13].

The aim of this study was to explore the prevalence of risk of malnutrition among persons in Swedish large, middle and small sized hospitals. In addition, the aim was to explore how well nutritional interventions are targeted towards patients at undernutrition risk.

Methods

Study Context

Swedish hospitals can be divided into regional hospitals, central general hospitals and general hospitals. Regional hospitals (or teaching hospitals, in this study labelled large sized hospitals with > 500 beds) have resources for the county, region, and in most cases also for high speciality national health care. Central general hospitals (in this study middle sized hospitals with 200–500 beds) are responsible for covering the whole county population's need for health care. The general hospitals (in this study small sized hospitals with < 200 beds) have the responsibility for a limited part of the county population's need for health care. The regional hospitals and the central general hospitals can provide 24-hour emergency admittance and have surgery departments, intensive care and radiology departments available on a 24-hour basis. These hospitals also have physicians within certain specialities (surgery, orthopaedics, medicine, gynaecology, radiology and anaesthesiology) on duty on a 24-hour basis.

Sample

The study was performed in 2007 during one single day in nine hospitals with a total coverage area of 1 197 500 inhabitants (Table 1).

The inclusion criteria were that all adult in-hospital patients (18 years or over) registered at the ward between 7 a.m. and 9 p.m. should be asked for participation. No intensive, out-patient or delivery units participated. Out of 2 620 patients, 2 170 (82.8%) agreed to participate. In large sized hospitals (n = 2) 1 197 (84.0%) out of 1 426 patients participated. In middle sized hospitals (n = 3) 669 (81.2%) out of 824 patients participated. In small sized hospitals (n = 4) 304 (82.2%) out of 370 patients participated. In the total sample, there were no significant differences regarding age (p-value = 0.086) and gender (p-value = 0.331) between those included (n = 2 167) and drop-outs (n = 445).

Table 1. Descriptions of included hospitals and the total coverage area of habitants.

	Hospital number	Coverage population
Large Sized Hospitals (n = 2)	1	274 000
Regional hospitals with > 500 beds	2	250 500
Middle Sized Hospitals (n = 3)	3	103 500
Central General Hospitals with 200–500 beds	4	163 500
	5	150 000
Small Sized Hospitals (n = 4)	6	93 000
General Hospitals < 200 beds	7	70 000
	8 and 9	93 000
Total Coverage Population		1 197 500

Procedure

Nursing students, clinical tutors and registered nurses and dieticians got training and education in how to collect the data and were then responsible for the data collection. Data was collected through measures of height and weight, interviews and observations of patients during mealtimes. The data collection was preceded by gaining informed consent.

Protocol and Definitions

The protocol contained three parts. The first part included background data about the patients. The second part included data about nutrition and eating difficulties. The third part contained information about nutritional support.

Risk of malnutrition in this study includes both undernutrition risk and overweight. Height and weight were measured using the standard equipment available at the particular units. Moderate/high undernutrition risk was defined as the occurrence of at least two of: involuntary weight loss, BMI below limit (BMI < 20 if </= 69 yrs, BMI < 22 if >/= 70 yrs), eating difficulties according to Minimal Eating Observation Form – Version II (MEOF-II) [13] based on Swedish recommendations for detecting undernutrition risk [6,14]. Information about unintentional weight loss was gained from the patient or estimated from previous weight.

Minimal Eating Observation Form—Version II (MEOF-II) includes three components of eating. Ingestion includes "manipulation of food on the plate," "transport of food to the mouth" and "sitting position." Deglutition includes "ability to chew," "manipulation of food in the mouth" and "swallowing." Energy includes "alertness," "appetite" and "eating < 3/4 of served food" [13].

Overweight was graded based on BMI (if </= 69 yrs: BMI =/> 25: if >/= 70 yrs: BMI =/> 27) and so was obesity (if </= 69 yrs: BMI 30–39: if >/= 70 yrs: BMI 32–41) and severe obesity (if </= 69 yrs: BMI =/> 40: if >/= 70 yrs: BMI =/> 42) [6].

Protein- and Energy Enriched food (PE-food) is food that is smaller in volume than the regularly served meals but has the same or higher content of protein and energy compared to the ordinary hospital food on the menu. "Supplements" include oral nutritional supplements such as protein and energy drinks given in addition to and chiefly between the main meals. Supplements do not include pharmacological therapy or drug supplement with multivitamin and mineral pills. Artificial nutrition includes enteral feeding (nasogastric tube, gastrostomy or jejunostomy) and parenteral feeding (via a peripheral or central vein) given alone or as a supplement to oral intake, and pre- and/or postoperative artificial nutrition with glucose or sodium chloride solutions. Eating assistance includes both partial (buttering bread, cutting food, only helping with beverage) and total assistance.

Analysis

Parametric and non-parametric statistics were used depending on the level of data and based on unpaired comparisons between two or three groups. The following tests were applied: Chi-square test, Kruskal Wallis test, Mann Whitney U-test, and one way ANOVA (with post-hoc analysis by Bonferroni correction). The level of statistical significance was set at P-value < 0.05. When multiple comparisons were made (going from three to two group comparisons) a reduced p-value of < 0.017 was used to avoid mass significance (type I or alpha error) [15]. Analyses were performed using SPSS version 16.0 (SPSS Inc., Chicago, IL, USA).

Ethics

The ethics for conducting scientific work was followed. This study was approved at each hospital. The patients were asked for informed consent. Both verbal and written information was given and patients were guaranteed anonymity, i.e. no personal identification number or names were collected. As the study was a part of an overall quality development project, no formal approval by an ethical committee was required, according to the Swedish Act concerning the Ethical Review of Research Involving Humans [16].

Results

There were significant differences between the three hospital samples regarding age. Patients in large sized hospitals were younger than those in middle and small sized hospitals, and those in middle sized hospitals were younger than those in small sized hospitals. Large and middle sized hospitals also had significantly more patients under surgical treatment compared to small sized hospitals. In addition there were significantly more psychiatric patients in small compared to in middle and large sized hospitals, and in middle compared to large sized hospitals. Correspondingly there were significant differences in diagnosis between the three samples. For instance, middle sized hospitals had more patients with gastrointestinal diseases, while the small sized had more patients with cardiovascular, psychiatric and orthopaedic diseases. There were significantly more patients with oncological diseases in large and middle sized than in small sized hospitals (Table 2).

Table 2. Characteristics of patients and divided according to the size of hospital.

| | Hospitals | | | |
	Large sized n = 1197	Middle sized n = 824	Small sized n = 370	P-value
Age, mean (SD)	66 (18)	69 (16)	70 (16)	0.001 a,b,c
Age group				< 0.005 a, b
< 70 years, %	51.7	45.0	39.9	
≥ 70 years, %	48.3	55.0	60.1	
Gender, %				0.580
Men	49.5	48.0	46.5	
Women	50.5	52.0	53.5	
Distribution of patients within some specialities, %				
Medicine	34.9	41.0	33.2	0.092
Surgery	16.8	17.0	10.5	0.022 a,c
Orthopaedics	8.6	11.1	13.8	0.018 d
Psychiatry	5.3	2.6	8.9	0.001 b,c
Distribution of patients according to some diagnosis categories, %				
Pulmonary	6.1	11.6	9.8	< 0.0005 a,b
Cardiovascular	19.0	22.4	30.6	< 0.0005 a,c
Infectious	9.9	7.3	4.0	0.003 b,c
Gastrointestinal	13.4	20.3	3.4	< 0.0005 a,b,c
Neurological	4.3	5.2	2.4	0.139
Orthopaedic	10.8	7.8	29.0	< 0.0005 a,b,c
Psychiatric	6.1	3.1	11.1	< 0.0005 a,b,c
Oncological	28.5	24.2	7.3	< 0.0005 b,c

Missing values in < 5%. ANOVA, Chi-square test
a first group differs compared to second
b first group differs compared to third
c second group differs compared to third group
d not significant in post hoc analysis

Eating difficulties were significantly more common among patients in large compared to middle and small sized hospitals, and in middle compared to small sized hospitals. Unintentional weight loss and undernutrition risk were significantly more common in large compared to middle and small sized hospitals. In large sized hospitals 38.2% of patients were overweight (including obesity) and in middle and small sized hospitals it was 42.6% and 42.2% respectively (no significant difference) (Table 3).

Table 3. Point prevalence of risk of undernutrition (UN) and overweight among the studied patients.

| | Hospitals | | | |
	Large sized n = 1197	Middle sized n = 824	Small sized n = 370	P-value
Criteria for UN-risk, %				
Eating difficulties according To MEOF-II	58.2	50.4	42.5	< 0.0005 a,b,c
Low BMI	22.3	19.9	17.2	0.145
Unintentional weight loss	40.2	34.1	27.2	0.001 a,b
Fulfilling UN risk criteria, %				< 0.0005 a,b
No criteria – no UN risk	31.0	35.2	47.3	
One criteria – low UN risk	35.0	38.6	31.1	
Two criteria – moderate UN risk	26.4	21.1	17.7	
Three criteria – high UN risk	7.6	5.1	3.9	
Overweight, %				0.102
No overweight	61.8	57.4	57.8	
Grade 1, overweight	26.0	27.0	31.0	
Grade 2, obesity	11.2	14.8	11.2	
Grade 3, severe obesity	1.0	0.8	0.0	

MEOF-II = Minimal Eating Observation Form – Version II. Missing values in < 10%. Chi-square test, Kruskal Wallis test, Mann Whitney U test
a first group differs compared to second
b first group differs compared to third
c second group differs compared to third group

Having ingestion and deglutition difficulties was significantly more common among patients in large compared to small sized hospitals. Energy problems were more common among patients in large sized compared to in middle and small sized hospitals, and in middle compared to small sized hospitals (Table 4).

Table 4. Distribution (%) of eating difficulties divided according to the size of each hospital.

	Large sized n = 1197	Hospitals Middle sized n = 824	Small sized n = 370	P-value
Eating difficulties				
Ingestion	12.2	9.0	8.0	< 0.029 b
Deglutition	20.5	14.0	15.3	0.001 b
Energy	50.3	43.5	34.2	< 0.0005 a,b,c

Missing values in < 5%. Chi-square test
a first group differs compared to second
b first group differs compared to third
c second group differs compared to third group

There were no significant differences between hospitals in the precision of nutritional interventions in relation to undernutrition risk despite for artificial feeding and in the provision of eating assistance. More patients at no/low risk of undernutrition got artificial nutrition in large compared to small sized hospitals and in middle compared to small sized hospitals. More patients in small sized hospitals were provided eating assistance compared to in middle and large sized hospitals. Among those with moderate/high risk of undernutrition, significantly more patients got artificial feeding in large sized and middle sized hospitals than in small sized hospitals (Table 5).

Table 5. The precision (%) in the nutritional care for patients at no/low risk and for patients at moderate/high risk of undernutrition.

	Large sized	Hospitals Middle sized	Small sized	P-value
At no/low risk of undernutrition	n = 760	n = 465	n = 222	
PE-food	5.3	4.0	1.9	0.122
Oral Supplement	16.3	19.0	13.5	0.102
Artificial nutritional (AN) support	15.0	11.3	3.3	< 0.0005 b,c
Pre- and/or postoperative AN	5.4	5.2	1.8	0.077
Eating Assistance	5.3	6.4	11.7	0.011 b,c
At moderate/high risk of undernutrition	n = 392	n = 165	n = 61	
PE-food	14.4	16.6	6.8	0.181
Oral Supplement	44.8	54.1	43.3	0.115
Artificial nutritional support	22.5	22.1	8.3	0.040 b,c
Pre- and/or postoperative AN	6.1	9.7	3.3	0.160
Eating Assistance	18.3	14.4	20.3	0.593

Missing values in < 15%. Chi-square test
a first group differs compared to second
b first group differs compared to third
c second group differs compared to third group

Patients with energy problems in small sized hospitals got less artificial feeding, less pre- and/or postoperative artificial nutrition and more eating assistance than patients in large and middle sized hospitals. Other than that, there were no significant differences between hospitals in the targeting of nutritional care in relation to energy problems (Table 6).

Table 6. The precision (%) in the nutritional care in relation to patients with energy problems (eat little, stop eating due to tiredness rather than to having satisfied their hunger, and eat slowly)

| | | Hospitals | | |
	Large sized	Middle sized	Small sized	P-value
Energy problems	n = 599	n = 276	n = 103	
PE-food	12.6	12.1	5.9	0.159
Oral Supplement	38.9	43.4	37.6	0.407
Artificial nutritional (AN) support	21.6	23.2	11.0	0.031 b,c
Pre- and/or postoperative AN	6.2	10.1	1.9	0.011 b,c
Eating Assistance	15.7	16.5	28.0	0.019 b,c

Missing values in < 15%. Chi-square test
a first group differs compared to second
b first group differs compared to third
c second group differs compared to third group

Discussion

The results of this study clearly demonstrate that there are significant differences in the prevalence of undernutrition risk in relation to hospital volume. Eating assistance is provided to a greater extent and artificial feeding to a lesser extent in small compared to in middle and large sized hospitals. Other than that, there is no difference in the precision of provision of nutritional care.

Prevalence studies always reflect a snapshot of reality and must therefore be interpreted with care. In this large survey many persons were involved in the data collection, which can be seen as a shortcoming of the study. However, all the staff responsible for data collection had got the same education about procedure, screening and how to fill in the study protocol. This method of data collection is very useful when the goal is to reach a large sample using limited resources. In addition, there are gains made for the students and clinical practitioners, such as awareness of research methodology and nutrition and eating difficulties, by involving staff and students in the data collection [17].

The same methodology used in the present study was used in an earlier study in 2005 [6] and the instrument MEOF II for detecting eating difficulties was then slightly modified based on psychometric criteria [13]. However, the combination of unintentional weight loss, low BMI and MEOF II for defining undernutrition risk need to be compared to other validated instruments in future studies.

The prevalence of undernutrition risk found in large sized hospitals cannot automatically be generalised to middle or small sized hospitals due to differences in patient populations. A stepwise decrease, from large sized to small sized hospitals, was found in the number of patients with moderate or high undernutrition risk. The same pattern was found in the British survey, with a higher prevalence of undernutrition risk in large sized hospitals than in smaller hospitals [9]. Such a pattern was expected (but not confirmed) in the German nationwide survey [10]. The researchers stated in the discussion that the prevalence of undernutrition risk was expected to be higher in the larger hospitals, as patients admitted to university hospitals might be more severely sick and thus more prone to malnutrition [10]. However, it has not been demonstrated that the patients (in general) in university hospitals are more sick than those in general hospitals. Instead, it can even be that teaching/university hospitals admit healthier patients than general hospitals and also perform higher volumes of procedures than general hospitals. At least, this seems to be the case in cardiology [18]. However, hypothetically it can be that the complexity of diseases rather than "severity of illness" [19] cause the higher prevalence of undernutrition risk. Also differences in comorbidity may explain the association between hospital volume and outcome, i.e. undernutrition [20]. The hypothesis about comorbidity/complexity/rarity is supported by the fact that more patients in large sized hospitals in the present study had different types of eating difficulties, and at the same time, they were younger than the patients in middle and large sized hospitals. In addition, it is known that advanced age predisposes to nutritional deficits [10]. If only age and not the characteristics of the case mix were considered as an explanation, one would expect a higher prevalence of undernutrition risk in smaller hospitals. However, in the present study there were many patients with oncological, gastrointestinal and cardiovascular diseases in the large sized hospitals, diagnoses known to involve a high prevalence of undernutrition risk [8,10]. It is difficult to draw any firm conclusions about the reasons for the higher prevalence of undernutrition risk in larger hospitals, but it is likely that the characteristics of the case mix in different hospitals is the cause of this phenomenon. There is a need to further explore the reasons behind the differences in prevalence of undernutrition risk in relation to hospital volume.

Significantly more patients in large and middle sized hospitals got artificial nutrition compared to in small hospitals, while patients in small sized hospitals got more assisted feeding. One explanation could again be that this difference reflects characteristics of the case mix in the different hospitals, or rather the adaptation of treatment due to specific disease characteristics only superficially controlled for in this study. A second and more controversial explanation could be that one is more prone to give artificial feeding to younger patients and feeding assistance to older patients. A third explanation could be that there is a culture in large sized hospitals to use technical solutions (artificial feeding) to a greater extent than in small sized hospitals. This last explanation is supported by the fact that university hospitals use "more procedures" (invasive investigations, treatments) than general hospitals [18,20]. However, in this study one should be careful in interpreting artificial feeding as an intervention decided on due to only undernutrition risk, as there could have been other reasons for this action. More studies are needed that explore the targeting of nutritional interventions towards those needing them most.

Larger hospitals have more patients with eating difficulties than smaller hospitals. Especially energy problems differed in relation to the size of hospital, with more patients having energy problems in large sized hospitals. A difference in the precision of common nutritional interventions (i.e. PE-food, oral supplements) could perhaps have been expected, by means of higher precision in the large sized and more specialised teaching hospitals. No other study has been found (PubMed search, January 2009) that looks specifically at the targeting of PE-food and oral supplements in relation to undernutrition risk. But a previous study [13] found that staffs are good at providing eating assistance for patients with ingestion difficulties, and that these problems do not strongly contribute to undernutrition risk. It has also been found that energy problems are the single most important factor among the eating difficulties that contribute to undernutrition risk [12,13]. A better targeting of PE-food and oral supplements is perhaps the most important step to take, as it is well known that dietary supplementation is beneficial by means of for instance better values in anthropometric measures, decreased hospital stay and mortality [21-26]. In this study, only 6–13% of patients with energy problems (eat little, stop eating due to tiredness, eat slowly) got PE-food and 39–43% of patients got oral supplements. Thus, it can be concluded that staff are less good at targeting these interventions (PE-food, supplements) towards patients with energy problems and that these problems therefore are likely to be among the strongest contributing factors to the development or maintenance of undernutrition risk [13].

Many patients admitted to hospitals are overweight. Between 38% and 43% were found to be overweight. Preventive actions such as information about the risks connected to overweight and the importance of exercise and eating healthy food, and help to overweight persons with losing weight, need to be taken, especially if there are weight-related health problems [27]. Studies have shown that weight-loss therapy improves physical functioning and quality of life, and decreases the medical complications associated with obesity in older persons [27]. However, voluntary weight loss needs to take place under controlled forms in order to not cause loss of bone mass or muscle mass, and not during the acute phase of disease. Thus, information should be given in hospitals and the weight-loss therapy can start after hospital discharge when the health status has stabilised. It is also important to combine weight-loss therapy with physical activities. To sum up, clinical practice needs to focus on both undernutrition and overweight.

Conclusion

The prevalence of undernutrition risk differs depending on the case mix that in turn is related to the hospital volume. There are no differences in the precision in providing PE-food and oral supplements that are due to the hospital volume, while there are differences in the type of eating difficulties that the patients have. It can be recommended that greater efforts should be taken to increase the use of PE-food and oral supplements, especially for patients suffering from a lack of energy (eat little, stop eating due to tiredness, poor appetite). Also great efforts need to be taken to decrease the occurrence of overweight. Thus the awareness among physicians, nurses and other professionals must be improved about how to increase the precision in provision of nutritional care. This can be done through education in nutritional screening, assessment and treatment and also by implementation of national recommendations in food and nutritional care in hospitals.

Competing Interests

The authors declare that they have no competing interests.

Authors' Contributions

AW was the lead investigator of the study (together with KU). He was involved in the design of the study, providing information to data collectors, analysing data and writing the manuscript. KU, CA, EBB, CWH, and JS were involved in the design of the study, providing information to data collectors and critically revising the manuscript. CL, RK and RS were involved in the design of the study and critically revising the manuscript. All authors besides KU (deceased) approved the final manuscript.

Acknowledgements

The first author is supported by the Swedish Research Council and the Skane county council's research and development foundation. We thank the patients and the staff, teachers and students for their cooperation. The study was conducted in cooperation with the clinical research group within the knowledge group for clinical nursing science at Kristianstad University College.

References

1. McWhirter JP, Pennington CR: Incidence and recognition of malnutrition in hospital. BMJ 1994, 308(6934):945–8.

2. Kondrup J, Johansen N, Plum LM, Bak L, Larsen IH, Martinsen A, Andersen JR, Baernthsen H, Bunch E, Lauesen N: Incidence of nutritional risk and causes of inadequate nutritional care in hospitals. Clin Nutr 2002, 21(6):461–8.

3. Rasmussen HH, Kondrup J, Staun M, Ladefoged K, Kristensen H, Wengler A: Prevalence of patients at nutritional risk in Danish hospitals. Clin Nutr 2004, 23(5):1009–15.

4. Council of Europe: Resolution ResAP (2003)3 on food and nutritional care in hospitals. [http://www.hospitalcaterers.org/documents/cu.pdf].

5. Johansson U, Larsson J, Rothenberg E, Stene C, Unosson M, Bosaeus I: [Nutritional care in hospitals. Swedish hospitals do not manage to follow the European committee's guidelines]. Lakartidningen 2006, 103(21–22):1718–20.

6. Westergren A, Lindholm C, Axelsson C, Ulander K: Prevalence of eating difficulties and malnutrition among persons within hospital care and special accommodations. J Nutr Health Aging 2008, 12(1):39–43.

7. Planas M, Audivert S, Pérez-Portabella C, Burgos R, Puiggrós C, Casanelles JM, Rosselló J: Nutritional status among adult patients admitted to an university-affiliated hospital in Spain at the time of genoma. Clin Nutr 2004, 23(5):1016–24.

8. Kruizenga HM, Wierdsma NJ, van Bokhorst MA, de van der Schueren , Haollander HJ, Jonkers-Schuitema CF, Heijden E, Melis GC, van Staveren WA: Screening of nutritional status in The Netherlands. Clin Nutr 2003, 22(2):147–52.

9. Russell CA, Elia M, on behalf of BAPEN and collaborators: Nutrition screening survey in the UK in 2007. Nutrition screening survey and audit of adults on admission to hospitals, care homes and mental health units. [http://www.bapen.org.uk/pdfs/nsw/nsw07_report.pdf]. BAPEN 2007.

10. Pirlich M, Schütz T, Norman K, Gastell S, Lübke HJ, Bischoff SC, Bolder U, Frieling T, Güldenzoph H, Hahn K, Jauch KW, Schindler K, Stein J, Volkert D, Weimann A, Werner H, Wolf C, Zürcher G, Bauer P, Lochs H: The German hospital malnutrition study. Clin Nutr 2006, 25(4):563–72.

11. Joosten E, Vanderelst B, Pelemans W: The effect of different diagnostic criteria on the prevalence of malnutrition in a hospitalized geriatric population. Aging 1999, 11(6):390–4.

12. Westergren A, Unosson M, Ohlsson O, Lorefält B, Hallberg IR: Eating difficulties, assisted eating and nutritional status in elderly (> 65 years) patients in hospital rehabilitation. Int J Nurs Stud 2002, 39(3):341–51.

13. Westergren A, Lindholm C, Mattsson A, Ulander K: Mimimal Eating Observation Form: Reliability and Validity. J Nutr Health Aging 2009, 13(1):6–12.

14. SNUS, Samarbetsgruppen för nutritionens utveckling i Sverige: [Cooperation group for the development of nutrition in Sweden] Nutritionsbehandling i sjukvård och omsorg. [Nutritional treatment in care and service]. Dietisternas Riksförbund, Nutritionsnätet för sjuksköterskor, Svensk Förening för Klinisk Nutrition, SWESPEN: Stockholm; 2004. (In Swedish)

15. Altman DG: Practical statistics for medical research. London: Chapman and Hall; 1991.

16. SFS 2003 460: Lag om etikprövning av forskning som avser människor [The Act concerning the Ethical Review of Research Involving Humans on the website of the Central Ethical Review Boards]. [http://www.sweden.gov.se/sb/d/3288/a/19569]

17. Ulander K, Westergren A, Axelsson L, Lindholm C: Building practice knowledge by assessing eating and malnutrition in a point prevalence study. [http://abstract.mci-group.com/cgi-bin/mc/dq.pl?ccode=ESPEN2006&show=TKNAS] ESPEN-conference, Istanbul 19–22 Oct,2006 2006, 141.

18. Cram P: Testimony of Peter Cram, assistant professor of medicine at the university of Iowa before the house subcomittee on health concerning speciality hospitals. [http:/ / archives.energycommerce.house.gov/ reparchives/ 108/ Hearings/ 05122005hearing1517/ Cram.pdf]. 2005.

19. Averill RF, McGuire TE, Manning BE, Fowler DA, Horn SD, Dickson PS, Coye MJ, Knowlton DL, Bender JA: A study of the relationship between severity of illness and hospital cost in New Jersey hospitals. Health Serv Res 1992, 27(5):587–606.

20. Halm EA, Lee C, Chassin MR: Is volume related to outcome in health care? A systematic review and methodologic critique of the literature. Ann Intern Med 2002, 137(6):511–20.

21. Olin AO, Osterberg P, Hadell K, Armyr I, Jerstrom S, Ljungqvist O: Energy-enriched hospital food to improve energy intake in elderly patients. JPEN J Parenter Enteral Nutr 1996, 20(2):93–7.

22. Turic A, Gordon KL, Craig D, Ataya DG, Voss AC: Nutrition supplementation enables elderly residents of long-term-care facilities to meet or exceed RDAs without displacing energy or nutrient intakes from meals. J Am Diet Assoc 1998, 98(12):1457–9.

23. Harris D, Haboubi N: Malnutrition screening in the elderly population. J R Soc Med 2005, 98(9):411–4.

24. Milne AC, Potter J, Avenell A: Protein and energy supplementation in elderly people at risk from malnutrition. Cochrane Database Syst Rev 2005, CD003288.

25. Milne AC, Avenell A, Potter J: Meta-analysis: protein and energy supplementation in older people. Ann Intern Med 2006, 144(1):37–48.

26. Labossiere R, Bernard MA: Nutritional considerations in institutionalized elders. Curr Opin Clin Nutr Metab Care 2008, 11(1):1–6.

27. Villareal DT, Apovian CM, Kushner RF, Klein S: Obesity in older adults: technical review and position statement of the American Society for Nutrition and NAASO, The Obesity Society. The American Journal of Clinical Nutrition 2005, 82(2):923–34.

Hospital Discharge Planning and Continuity of Care for Aged People in an Italian Local Health Unit: Does the Care-Home Model Reduce Hospital Readmission and Mortality Rates?

Gianfranco Damiani, Bruno Federico, Antonella Venditti, Lorella Sicuro, Silvia Rinaldi, Franco Cirio, Cristiana Pregno and Walter Ricciardi

ABSTRACT

Background

Hospital discharge planning is aimed to decrease length of stay in hospitals as well as to ensure continuity of health care after being discharged.

Hospitalized patients in Turin, Italy, who are in need of medical, social and rehabilitative care are proposed as candidates to either discharge planning relying on a care-home model (DPCH) for a period of about 30 days, or routine discharge care. The aim of this study was to evaluate whether a hospital DPCH that was compared with routine care, improved patients' outcomes in terms of reduced hospital readmission and mortality rates in patients aged 64 years and older.

Methods

In a retrospective observational cohort study a sample of 380 subjects aged 64 years and over was examined. Participants were discharged from the hospital S.Giovanni Bosco in Turin, Italy from March 1st, 2005 to February 28th, 2006. Of these subjects, 107 received routine discharge care while 273 patients were referred to care-home (among them, 99 received a long-term care intervention (LTCI) afterwards while 174 did not). Data was gathered from various administrative and electronic databases. Cox regression models were used to evaluate factors associated with mortality and hospital readmission.

Results

When socio-demographic factors, underlying disease and disability were taken into account, DPCH decreased mortality rates only if it was followed by a LTCI: compared to routine care, the Hazard Ratio (HR) of death was 0.36 (95% Confidence Interval (CI): 0.20 – 0.66) and 1.15 (95%CI: 0.77 – 1.74) for DPCH followed by LTCI and DPCH not followed by LTCI, respectively. On the other hand, readmission rates did not significantly differ among DPCH and routine care, irrespective of the implementation of a LTCI: HRs of hospital readmission were 1.01 (95%CI: 0.48 – 2.24) and 1.18 (95%CI: 0.71 – 1.96), respectively.

Conclusion

The use of DPCH after hospital discharge reduced mortality rates, but only when it was followed by a long-term health care plan, thus ensuring continuity of care for elderly participants.

Background

Intermediate care is aimed to facilitate transition from hospital to home when the objectives of care are not primarily medical: patients are discharged earlier, and hospital length of stay is decreased [1]. In line with the principle of 'care closer to home,' intermediate care services should generally be provided in community-based settings, in the patient's home, or they may be provided in discrete step-down facilities on acute hospital sites [2].

The need for effective discharge planning for elderly patients is becoming increasingly important due to the rising number of elderly people requiring hospital care, pressure on beds and recognition of the problems surrounding hospital discharge [3]. The main problems concern poor communication between hospital and community [4-8], lack of assessment and planning for discharge [4,9] and inadequate notice of discharge to the patients [9,10]. Furthermore, discussion of discharge with patients and their caregivers has been generally infrequent [10,11]. Over-reliance on informal support and/or poor statutory service provision [10,12-14], lack of attention to the individual needs of the most vulnerable [9,15], and wasted or duplicated visits by community nurses [5] were reported in the literature.

Discharge planning aims to review current medication, facilitate compliance with established treatment, improve home functioning and safety, prevent unnecessary hospital admission, and promote effective rehabilitation services. It also aims to enable early discharge from hospital and prevent premature or unnecessary admission to long-term residential care. In Italy, this form of discharge is put into effect in Local Health Units, public enterprises which are legally creatures of the regions with administrative and financial independence [16,17]. In the Piedmont region, where the local population is 4,250,775 and the number of elderly is about 900,000 (21.2%), the problem of continuity of care for aged patients is particularly relevant [18]. In each Local Health Unit of Piedmont, an Operative Care Centre aims at managing community hospital, residential and home care services after hospital discharge of these patients [19]. Patient problems during post-discharge may vary over time and are often accompanied by unmet needs. These problems may be related to their physical, functional, emotional and social status and include patient-related factors, care-related factors and features related to the social network of the patient. The patient-related problems are associated with a decline in physical health status (such as physical complaints), with decreased functional status (difficulty in performing the activities of daily life, and/ or need of care with these activities), or with disturbed emotional status (feeling insufficiently informed or having uncertain, negative feelings and emotional worries). Examples of health care related factors are the way in which patients and their family are prepared for discharge and the post-discharge period, home care is provided, and the extent to which hospital and home-care are inter-related. Finally, features related to the social network are the availability, the skills and the willingness of the social network to provide support and/or help for the patients. In all these areas patients might experience insufficient or inadequate support in coping with the difficulties or limitations involved, which can sometimes result in hospital readmissions. Problems after discharge and the influencing factors are interrelated, in concept and over time. Furthermore, the literature shows that post-discharge problems can be reduced by efficient discharge planning during hospitalization and by intensive after care, and that the risk of post-discharge problems can be predicted to some extent [20].

So far, most research has focused on the effectiveness of hospital discharge planning based on interventions delivered at home that are compared to routine care. Among patients with hip fractures discharged from a medical centre in northern Taiwan, it was found that patients who received hospital discharge planning had a shorter length of stay, lower rate of readmission, and higher survival rate compared to those who received routine care [21]. Phillips et al.'s meta-analysis indicated that patients undergoing post-hospital discharge planning had lower mortality or readmission rate (for the combined end-point, Relative Risk (RR) = 0.73; 95% CI 0.62–0.87) [22] compared to those receiving routine care. In the Parker et al. review, readmission's RR was 0.85 (95% CI, 0.76–0.95), indicating a reduction in relative risk of being readmitted for patients receiving post-hospital discharge protocols [23].

On the basis of fifteen reviews, Mistiaen et al concluded that there is only limited evidence for the positive impact on readmission rates of discharge interventions. Discharge interventions did not appear to be effective for three reviews in which the largest effects were observed when interventions from the discharge planning and discharge support side were combined across the hospital-home interface. In addition, two reviews showed that educational interventions might have some effect on aspects of the emotional status after discharge, on knowledge and medication adherence [20]. The limited evidence about effectiveness of discharge interventions may be due to the heterogeneity of studies. In addition, Shepperd et al. found that there were no statistically significant differences in mortality (Odds Ratio (OR) = 1.44, 95% CI 0.82–2.51) and readmission between discharge planning and routine discharge care (OR = 0.91, 95% CI 0.67–1.23) [24]. There was, however, some evidence that services combining needs assessment, discharge planning and a method for facilitating the implementation of these plans were more effective than services that do not include the latter action [25].

These findings indicate that discharge planning is likely to play a key role as a management tool in intermediate care when the latter is provided at patient's home. The National Health Service in the UK has recently commissioned an evaluation of intermediate care for older people: current evidence suggested substantial changes in service organization and provision, and favourable experience reported by the users [26].

However, since there is a wide diversity of provision of services and the lack of a standard terminology for what constitutes recovery or rehabilitation, there have been relatively few studies that specifically focus on the effectiveness of intermediate care in residential settings. As pointed out by Plochg et al., the setting up of intermediate care may encounter several difficulties [27]. No significant differences in mortality or hospital re-admission were found for subjects who were treated in an intermediate residential setting for 6 weeks compared to those who received routine care at home [28]. Consequently, the aim of this study was to evaluate whether a hospital discharge planning in a care-home setting (DPCH), compared with routine care, improved patient's outcomes. More specifically, our aim was to evaluate the effectiveness of DPCH in terms of reduced hospital readmission and mortality rates in patients aged 64 years and over in one of the main Local Health Unit of Piedmont, Italy.

Methods

Study Design

In this retrospective observational cohort study, we focused on a sample of 380 subjects aged 64 years and older who were discharged from the hospital "S. Giovanni Bosco" in Turin, Italy. Among these patients, 273 received a hospital discharge planning in a DPCH, while 107 patients received routine discharge care. We included in the study patients discharged between March 1st, 2005 and February 28th, 2006, whose age at hospital discharge was 64 years and over. All patients were discharged alive from the hospital and they were observed for a minimum of six months.

Before hospital discharge, a team composed of a geriatrician, a district nurse and social workers determined the medical, psychological, and functional capabilities of the elderly person in order to develop an integrated plan for treatment and follow-up after hospital discharge [29-32]. The evaluation was carried out through the use of instrumental scales (Activity of Daily Living – ADL, Instrumental ADL – IADL, Short Portable Mental Status Questionnaire – SPMSQ) and it was needed to identify the level of complexity of care. In addition, an evaluation of the presence of care givers, family network, presence of voluntary association and housing conditions lead to the definition of the appropriate social care. After the need of medical, social and rehabilitative treatments were taken into account, patients were assigned to receive either DPCH for about 30 days [19] or routine care at home. Patients were referred to DPCH mainly when there was the need of monitoring the effect of new prescribed therapies and/or they needed physical rehabilitation. The care-home setting consisted of 2 residential homes, with a total of 43 beds.

For subjects who received DPCH, individualised care pathways were provided by a multidisciplinary team involving nurses, physical therapists, occupational therapists, geriatricians, community care officers and social workers on a 24-hour basis. A nurse "case manager" was in charge of patient safety and monitored the implementation of the care plan, with the aims of improving patients' level of autonomy and supporting the creation of an adequate care network. Physical and occupational therapists were mostly engaged in developing patients' skills for daily living activities. After a period of about 30 days, a further Operative Care Centre assessment was performed. The pre-post comparison of scores for DPCH patients showed a slight improvement, especially for IADL scores, although this was not statistically significant at the 0.05 level. Patients could then be entitled to receive a long-term care intervention (LTCI) within the same residential setting. This included health and social interventions, which were carried out without a pre-defined duration in time, mainly by nurses, community care officers and social workers. These interventions supported individuals in the activities of daily living. They were provided on the basis of an individual plan, implemented by the same multidisciplinary team, and managed by the same nurse "case manager" that had intervened in the intermediate phase.

In the case of routine care, patients were discharged from hospital to home after the needs assessment, and received the usual health and social care they would ordinarily receive. At home, they were periodically visited by their general practitioner, nurses, physiotherapists, geriatricians, community care officers and social workers. Patients received nursing interventions of varying levels of complexity and frequency, and the appropriate social care, without the coordination of a specific nurse "case manager." When required, medical specialists provided their services. In case of palliative care a nurse specialized in palliative treatments was added to the nursing team.

Data Sources

Data were extracted from different electronic databases that included:

- Hospital discharge records of S. G. Bosco Hospital containing International Classification of Diseases IX-Clinical Modification (ICD IX-CM) pathology codes and readmissions date;
- Data on discharge planning of the Operative Care Centre of Local Health Unit N°4 of Turin containing demographic variables (gender, age, etc) as well as physical and mental disability scales, such as ADL, IADL, and SPMSQ;
- Data from the registrar's office of the municipality of Turin in order to verify deaths and date of event;
- Data from the social services of the municipality of Turin containing social variables (family network, pension, etc).

Information on socio-demographic characteristics was collected for the following variables: gender, age at hospital admission (64–74, 75–84, 85+), living arrangements (living alone, living with at least a relative, caregiver), pension (< 750 euro, ≥ 750). Information on care needs, evaluated before hospital discharge, was categorised as follows: ADL scale (independent, partially dependent, and heavily dependent), IADL scale (independent, partially dependent, heavily dependent), and SPMSQ scale cognitive deterioration (absent-light, moderate and severe). The main reason for hospitalization was coded using the ICD-IX CM and then categorised according to the Major Disease Category in cardio-circulatory diseases, injury and poisonings, cancer, diseases of the respiratory system, and diseases of the digestive system. A further category was created for diseases not included in the previous Major Disease Category. The prescription and implementation of LTCI were derived from Operative Care Centre archives.

Statistical Analyses

Descriptive and inferential analyses were performed using SPSS 13.0. All subjects were followed up for a minimum of 6 months. Log-rank test with significance level of alpha = 0.05 was used to evaluate associations between type of care and each dependent variable (mortality and readmission) over the follow-up period. Two separate Cox regression analyses were applied to estimate adjusted Hazard Ratio (HR) with 95% Confidence Interval (95% CI) of death and hospital readmission, respectively. The variables that were significant at the univariate analysis at the alpha = 0.20 level were included in the Cox regression models. The p-value of log partial ratio test was evaluated to assess the significance of fitted models. Assumptions of hazards proportionality for Cox regression model were checked by Schoenfeld residuals and Log-Minus-Log plots.

Since the follow-up period was longer than the duration of stay in DPCH (i.e. 30 days at maximum), we took into account a relevant factor intervening after this period, which might have affected our outcome measures, that is the implementation of a LTCI plan. Therefore, the comparison is not limited to two groups (i.e. DPCH vs. routine care) but instead it is made among three groups (two subgroups of DPCH, according to the implementation of a long-term care plan during the follow-up period, vs. routine care).

Ethics

Approval of the ethics committee was not required for the study. Data were extracted from routinely collected administrative databases and there was no need to obtain additional data from individual patients. The interventions under study were performed in ordinary or "natural" conditions, irrespective from the conduct of the present study. Because this was an observational retrospective study, patients had already been treated when the study protocol was written. Data linkage was performed by the team directly involved in patients' care using numerical codes. For the present study, researchers had access only to an anonymous dataset, which ensured patients' privacy. For these reasons, no personal informed consent to the present analysis was requested from study participants.

Results

The socio-demographic characteristics of patients are presented in Table 1. This Table shows the number and percentages of subjects according to socio-demographic variables, reason for hospital admission and functional status, among the three study groups. In the overwhelming majority of subjects (94.4%) who received routine care no LTCI was implemented. The majority of subjects were elderly aged 75–84 years: 36.4% of those who received routine care, 42.4% in those admitted to care-home followed by implementation of long term care, and 53.4% among subjects admitted to care-home not followed by implementation of long-term care. There were statistically significant differences in the main reason for hospitalization among the three groups (p < 0.001); cancer was more frequent among those who received routine care than among both subgroups of DPCH (29.0% vs. 9.1% and 8.0%), while injuries were dominant in DPCH (6.5% vs. 19.2% and 25.3%).

Table 1. Number and percentage of typology of intermediate care and characteristics of the subjects

		DPCH		
	Routine care	Followed by LTCI	Not followed by LTCI	p-value (Chi-square test)
	n = 107	n = 99	n = 174	
Gender				
Female	59 (55.1%)	66 (66.7%)	108 (62.1%)	0.238
Male	48 (44.9%)	33 (33.3%)	66 (37.9%)	
Age (years)				
64–74	34 (31.8%)	30 (30.3%)	23 (13.2%)	0.003
75–84	39 (36.4%)	42 (42.4%)	93 (53.4%)	
85+	34 (31.8%)	27 (37.4%)	58 (33.3%)	
Living arrangement				
Living alone	44 (41.1%)	51 (51.5%)	91 (52.3%)	0.427
Living with at least a relative	48 (44.9%)	35 (35.4%)	63 (36.2%)	
Caregiver	15 (14.0%)	13 (13.1%)	20 (11.5%)	
Pension				
<750 euro	63 (58.9%)	51 (51.5%)	113 (65.3%)	0.08
≥750 euro	44 (41.1%)	48 (48.5%)	60 (34.7%)	
Primary diagnosis at admission				
Cardio-circulatory diseases	26 (24.3%)	36 (36.3%)	46 (26.4%)	<0.001
Injury and poisonings	7 (6.5%)	19 (19.2%)	44 (25.3%)	
Cancer	31 (29.0%)	9 (9.1%)	14 (8.0%)	
Diseases of the respiratory system	9 (8.4%)	10 (10.1%)	21 (12.1%)	
Diseases of the digestive system	6 (5.6%)	7 (7.1%)	14 (8.0%)	
Other diseases	28 (26.2%)	28 (28.3%)	35 (20.1%)	
ADL at hospital admission				
Independent	10 (9.3%)	15 (15.2%)	12 (7.0%)	<0.001
Partially dependent	55 (51.4%)	79 (80.8%)	135 (78.9%)	
Totally dependent	42 (39.3%)	4 (4.1%)	24 (14.0%)	
IADL at hospital admission				
Independent	6 (5.6%)	13 (13.3%)	23 (13.4%)	0.001
Partially dependent	33 (30.8%)	50 (51.0%)	57 (33.1%)	
Totally dependent	68 (63.6%)	35 (35.7%)	92 (53.5%)	
Cognitive deterioration				
Absent-light	76 (71.0%)	79 (79.8%)	127 (73.4%)	0.165
Moderate	22 (20.6%)	18 (18.2%)	29 (16.8%)	
Severe	9 (8.4%)	2 (2.0%)	17 (9.8%)	

DPCH: discharge planning relying on a care-home model.
LTCI: long term care intervention.
ADL: activity of daily living.
IADL: instrumental activity of daily living.

As shown in Table 2, patients receiving routine care had higher crude mortality rates than those in routine care (45.8% vs. 10.1% and 22.4%, p < 0.001) after six months of follow-up. The difference was especially marked in the case of cancer (77.4% vs. 22.2% and 50.0%, p = 0.007) and cardio-circulatory diseases (42.3% vs. 7.7% and 13.0%, p = 0.002).

Table 2. Number and Percentage of deaths after six months of follow-up according to intermediate care typology and characteristics of subjects

	Routine care	DPCH Followed by LTCI	DPCH Not followed by LTCI	p-value (Log-Rank test*)
	49 (45.8%)	10 (10.1%)	39 (22.4%)	
Gender				
Female	22 (37.3%)	4 (6.1%)	18 (16.7%)	0.002
Male	27 (56.3%)	6 (18.2%)	21 (31.8%)	
Age (years)				
64–74	18 (52.9%)	1 (5.0%)	3 (13.0%)	0.488
75–84	18 (46.2%)	4 (9.5%)	19 (20.4%)	
85+	13 (38.2%)	5 (13.5%)	17 (29.3%)	
Living arrangement				
Living alone	20 (45.5%)	5 (9.8%)	17 (18.7%)	0.570
Living with at least a relative	25 (52.1%)	3 (8.6%)	17 (27.0%)	
Caregiver	4 (26.7%)	2 (15.4%)	5 (25.0%)	
Pension				
<750 euro	28 (44.4%)	4 (7.8%)	26 (23.0%)	0.751
≥ 750 euro	21 (47.7%)	6 (12.5%)	13 (21.7%)	
Primary diagnosis at admission				
Cardio-circulatory diseases	11 (42.3%)	2 (7.7%)	6 (13.0%)	<0.001
Injury and poisonings	0 (0.0%)	1 (5.3%)	8 (18.2%)	
Cancers	24 (77.4%)	2 (22.2%)	7 (50.0%)	
Diseases of the respiratory system	3 (33.3%)	1 (10.0%)	8 (38.1%)	
Diseases of the digestive system	2 (33.3%)	1 (14.3%)	2 (14.3%)	
Other diseases	9 (32.1%)	3 (10.7%)	8 (22.9%)	
ADL at hospital admission				
Independent	4 (40.0%)	1 (6.7%)	3 (25.0%)	0.030
Partially dependent	26 (47.3%)	7 (8.9%)	27 (20.0%)	
Totally dependent	19 (45.2%)	1 (25%)	9 (37.5%)	
IADL at hospital admission				
Independent	1 (16.7%)	1 (7.7%)	3 (13.0%)	0.324
Partially dependent	21 (63.6%)	3 (6.0%)	15 (26.3%)	
Totally dependent	27 (39.7%)	5 (14.3%)	21 (22.8%)	
Cognitive deterioration				
Absent-light	34 (44.7%)	8 (10.1%)	28 (22.0%)	0.603
Moderate	13 (59.1%)	2 (11.1%)	6 (30.7%)	
Severe	2 (22.2%)	0 (0.0%)	5 (29.4%)	

DPCH: discharge planning relying on a care-home model
LTCI: long term care intervention
ADL: activity of daily living
IADL: instrumental activity of daily living
* Log-Rank test was applied within the whole follow-up period.

Table 3 shows the crude hospital readmission rates after six months of follow up for the three groups. About one in 5 subjects was re-admitted to the hospital within 6 months. Readmission was lower for subjects discharged to care-home (22.2% and 19.0% vs. 27.1%), but the difference was not statistically significant.

Table 3. Number and Percentage of Readmissions after six months of follow-up according to intermediate care typology and characteristics of subjects

	Routine care	DPCH Followed by LTCI	DPCH Not followed by LTCI	p-value (Log-Rank test*)
	29 (27.1%)	22 (22.2%)	33 (19.0%)	
Gender				
Female	17 (28.8%)	11 (16.7%)	18 (16.7%)	0.129
Male	12 (25.0%)	11 (33.3%)	15 (22.7%)	
Age (years)				
64–74	8 (23.5%)	4 (20.0%)	2 (8.7%)	0.074
75–84	12 (30.8%)	12 (28.6%)	21 (22.6%)	
85+	9 (26.5%)	6 (16.2%)	10 (17.2%)	
Living arrangement				
Living alone	16 (36.4%)	10 (19.6%)	17 (18.7%)	0.393
Living with at least a relative	9 (18.8%)	9 (25.7%)	12 (19.0%)	
Caregiver	4 (26.7%)	3 (23.1%)	4 (20.0%)	
Pension				
<750 euro	14 (22.2%)	11 (21.6%)	23 (20.4%)	0.472
≥ 750 euro	15 (34.1%)	11 (22.9%)	10 (16.7%)	
Primary diagnosis at admission				
Cardio-circulatory diseases	11 (42.3%)	4 (15.4%)	7 (15.2%)	0.256
Injury and poisonings	3 (42.9%)	4 (21.1%)	2 (4.5%)	
Cancers	6 (19.4%)	0 (0.0%)	4 (28.6%)	
Diseases of the respiratory system	3 (33.3%)	4 (40.0%)	7 (33.3%)	
Diseases of the digestive system	1 (16.7%)	3 (42.9%)	4 (28.6%)	
Other diseases	5 (17.9%)	7 (25.0%)	9 (25.7%)	
ADL at hospital admission				
Independent	3 (30.0%)	2 (13.3%)	2 (16.7%)	0.360
Partially dependent	17 (30.9%)	19 (24.1%)	27 (20.0%)	
Totally dependent	9 (21.4%)	1 (25.0%)	4 (16.7%)	
IADL at hospital admission				
Independent	2 (33.3%)	4 (30.8%)	2 (8.7%)	0.162
Partially dependent	8 (24.2%)	11 (22.0%)	17 (29.8%)	
Totally dependent	19 (27.9%)	6 (17.1%)	14 (15.2%)	
Cognitive deterioration				
Absent-light	20 (26.3)	15 (19.0%)	24 (18.9%)	0.103
Moderate	7 (31.8)	6 (33.3%)	4 (13.8%)	
Severe	2 (22.2)	1 (50.0%)	5 (29.4%)	

DPCH: discharge planning relying on a care-home model
LTCI: long term care intervention
ADL: activity of daily living
IADL: instrumental activity of daily living
*Log-Rank test was applied within the whole follow-up period.

The results of the multivariable Cox regression analyses are shown in tables 4 and 5. The independent predictors of mortality (Table 4) were cancer (HR = 3.27, 95% CI 1.93 – 5.55), diseases of respiratory system (HR = 1.84, 95% CI: 1.01 – 3.34), while DPCH followed by LTCI significantly decreased mortality compared to routine care (HR = 0.36; 95% CI 0.20–0.66). No significant difference was found between routine care and DPCH, when this was not followed by LTCI (HR = 1.15 95%CI: 0.77 – 1.74).

Table 5 shows HR and 95% CI for hospital readmission. Having a severe cognitive deterioration was a risk factor for readmission (HR = 2.20; 95% CI: 1.09–4.43). Both subgroups of DPCH showed similar hazards of readmission compared to routine care: HRs of hospital readmission were 1.01 (95%CI: 0.48 – 2.24) and 1.18 (95%CI: 0.71 – 1.96), for DPCH followed by LTCI and DPCH not followed by LTCI, respectively.

Table 4. Hazard Ratio of death

	HR	95% CI
Gender		
Female	1	
Male	1.44	1.00–2.09
Age (years)		
64–74	1	
75–84	0.77	0.48–1.26
85+	1.12	0.69–1.83
Primary diagnosis at admission		
Cardio-circulatory diseases	1	
Injury and poisonings	0.62	0.31–1.25
Cancers	3.27	1.93–5.55
Diseases of the respiratory system	1.84	1.01–3.34
Diseases of the digestive system	1.27	0.54–2.99
Other diseases	0.94	0.55–1.63
ADL at hospital admission		
Independent	1	
Partially dependent	0.57	0.26–1.24
Totally dependent	0.7	0.30–1.64
IADL at hospital admission		
Independent	1	
Partially dependent and Totally dependent	2.02	0.79–5.21
Typology of term–care		
Routine care	1	
Care home not followed by long term care intervention	1.15	0.77–1.74
Care home followed by long term care intervention	0.36	0.20–0.66

P-value of log partial likelihood ratio test <0.001

HR; hazard ratio
CI; confidence interval
ADL; activity of daily living
IADL; instrumental activity of daily living.

Table 5. Hazard Ratio of hospital readmission

	HR	95% CI
Gender		
Female	1	
Male	1.43	0.94–2.16
Age (years)		
64–74	1	
75–84	1.4	0.81–2.40
85+	0.9	0.49–1.65
Living arrangement		
Living alone	1	
Living with at least a relative or caregiver	0.72	0.48–1.10
Primary diagnosis at admission		
Cardio-circulatory diseases	1	
Injury and poisonings	0.83	0.43–1.62
Cancers	0.99	0.49–2.00
Diseases of the respiratory and digestive systems	1.69	0.97–2.96
Other diseases	0.99	0.57–1.72
IADL at hospital admission		
Independent	1	
Partially dependent	1.47	0.70–3.07
Totally dependent	1.01	0.48–2.14
Cognitive deterioration		
Absent-light	1	
Moderate	1.31	0.78–2.17
Severe	2.2	1.09–4.43
Typology of term-care		
Routine care	1	
Care home not followed by long term care intervention	1.18	0.71–1.96
Care home followed by long term care intervention	1.01	0.48–2.14

P-value of log partial likelihood ratio test = 0.049

HR; hazard ratio
CI; confidence interval
IADL; instrumental activity of daily living.

Discussion

In our study, when socio-demographic factors, underlying disease and disability were taken into account, hospital discharge planning implemented in a residential care-home setting decreased mortality rates only if it was followed by a LTCI. On the other hand, adjusting for socio-demographic characteristics, health and functional status, readmission rates did not significantly differ among DPCH and routine care, irrespectively of the implementation of a LTCI.

About one in 5 patients (22.1%) was readmitted within 6 months from hospital discharge. This re-admission rate is aligned with that of Trappes-Lomax et al. [28]. To the best of our knowledge, no published study has so far assessed the effectiveness of DPCH taking into account the implementation of subsequent long-term care plans according to this logic of continuity of care. In this context, we chose an observational study design which is a very practical and useful research tool, given the complexity of the scenario. Consequently, our study may suffer from the typical limitations of observational studies, that is the incomparability of groups: since subjects were referred to the different types of care according to clinical judgements, social and organizational matters, systematic differences may have occurred in baseline characteristics of subjects. However, in the phase of data-analysis, we took into account some of the major confounders, that are socio-demographic, clinical and functional characteristics, by means of a multivariable regression model. The high death rate (77.4%) observed within 6 months from hospital discharge among cancer patients who received routine care may contribute to the different mortality experience of patients. However, similar results were found at stratified analyses. We also performed a multivariable regression analysis excluding cancer patients. The results confirmed the protective effect of DPCH when this was followed by a LTCI, with HR = 0.43 (95%CI: 0.22, 0.84).

Previous studies comparing residential care home intermediate services and routine care did not show differences on mortality and readmission rates after hospital discharge among elderly subjects. In one study, subjects who were referred to a care home rehabilitation service did not show reduced hospital readmission rates after 3 and 12 months of follow-up [33]. Similarly, no significant differences in the hazard of hospital readmission or death was found after 6 and 12 months of follow-up between a joint health/social care residential rehabilitation unit and "usual" care in the UK [28]. However, these studies do not report any information on any LTCI which the subjects may have received over the follow-up period. Lack of information on LTCI implemented over the follow-up and complementary with intermediate care services, does not permit a complete assessment of the appropriateness of the continuity of care. According to a recent review, continuity of care has two key elements: care of an individual patient and care delivered over time [34]. In particular, "management continuity" plays an important role especially in chronic or complex clinical diseases that require management from several providers. Shared management plans and care protocols facilitate management continuity, providing predictability and security in future care for both patients and providers. Therefore, in order to be maximally effective, "management continuity" should be planned systematically in advance, involving all the relevant actors in both interfaces of care. The first interface is the outward hospital interface, which is the transition from hospital to residential intermediate care-home services, while the second one refers to the transition from intermediate to long-term care.

Even when long-term care is deemed necessary, many different reasons may hinder its implementation, especially in the home setting. In our study, these were mainly related to financial difficulties, such as co-payment of social services, and organisational problems, such as delays in performing the multidimensional assessment, the existence of waiting lists for residential services and delays in the provision of home care. The availability of both intermediate care-home and long-term care services within the same facility, as shown in our study, might help overcome the aforementioned difficulties and create more confidence in patient and care givers. This is supported by the finding that proposed interventions of long-term care were more frequently provided if patients were referred to DPCH in comparison to routine care. In addition, this organisational formula may determine a better efficiency in the use of health care resources.

Conclusion

The management of the continuity of care is fundamental especially in chronic or complex clinical diseases that require the contribution of several providers and personnel and are often implemented in different settings. Timely and shared plans ruling both the transitions between hospital and intermediate care, and between intermediate and long term care may determine better patients' outcomes. In this study, we attempted to open the "black box" of intermediate care, by describing context, setting and staffing involved. Future studies should focus on the evaluation of the effectiveness of hospital discharge planning taking into account the implementation of long-term care services.

Abbreviations

DPCH: Discharge Planning relying on a Care-Home model; LTCI: Long Term Care Intervention; HR: Hazard Ratio; CI: Confidence Interval; ADL: Activity of Daily Living; IADL: Instrumental Activity of Daily Living; SPSMQ: Short Portable Mental Status Questionnaire

Competing Interests

The authors declare that they have no competing interests.

Authors' Contributions

All authors contributed to the conception of this paper, and to the acquisition of data. GD wrote the first draft and all authors made important contributions to subsequent drafts. All authors have seen and approved the final version. GD and BF had full access to all of the data in the study and take responsibility for the integrity of the data and the accuracy of the data analysis.

Acknowledgements

We wish to thank Khairoonisa Foflonker for revising the manuscript.

References

1. Melis RJF, Olde Rikkert MGM, Parker SG, van Eijken MIJ: What is intermediate care? BMJ 2004, 329:360–361.

2. Fernandes U, Crane G, Butt AS, Patel CJ: Intermediate Care in Brent. Report of the Scrutiny Task Group Final Report. London 2005.

3. Department of Health: The NHS plan. A plan for investment, a plan for reform. London: Stationery Office; 2000.

4. Department of Health: The national service framework for older people. London: DoH; 2001.

5. Martin GP, Hewitt GJ, Faulkner TA, Parker H: The organization, form and function of intermediate care services and systems in England: results from a national survey. Health Soc Care Community 2007, 15:146–54.

6. Closs SJ, Tierney AJ: The complexities of using a structure, process and outcome framework: the case of an evaluation of discharge planning for elderly patients. J Adv Nurs 1993, 18:1279–87.

7. Skeet M: Home From Hospital The Result of a Survey Conducted Among Recently Discharged Hospital Patients. London: Macmillan Journals; 1970.

8. Armitage S: Liaison and Continuity of Nursing Care Executive Summary. Cardiff: Welsh Office; 1990.

9. Curran P, Gilmore DH, Beringer TRO: Communication of discharge information for elderly patients in hospital. The Ulster Medical Journal 1992, 61: 56–58.

10. Meara JR, Wood JL, Wilson MA, Hart MC: Home from hospital a survey of hospital discharge arrangements in Northamptonshire. Journal of Public Health Medicine 1992, 14:145–150.

11. Williams EI, Greenwell J, Groom LM: The care of people over 75 years old after discharge from hospital an evaluation of timetabled visiting by health visitor assistants. Journal of Public Health Medicine 1992, 14:138–144.

12. Harding J, Modell M: Elderly peoples' experiences of discharge from hospital. Journal of the Royal College of General Practitioners 1989, 39:17–20.

13. Victor C, Vetter NJ: Preparing the elderly for discharge from hospital a neglected aspect of patient care? Age and Ageing 1988, 17:155–163.

14. Congdon JG: Management the incongruities an analysis of hospital discharge of the elderly. Communicating Nursing Research, Nursing Research-Transcending the 20th Century 1990, 23:9–17.

15. Klop R, van Wijmen FCB, Philipsen H: Patients' rights and the admission and discharge process. Journal of Advanced Nursing 1991, 16:408–412.

16. France G, Taroni F, Donatini A: The Italian health-care system. Health Econ 2005, 14:S187–S202.

17. Jommi C, Cantu' E, Anessi-Pessina E: New funding arrangements in the Italian National Health Service. Int J Health Plann Manage 2001, 16:347–368.

18. Istat: Health for All – Italia. Un sistema informativo territoriale su sanità e salute. [http://www.istat.it/sanita/Health/] 2005.

19. D.G.R. n. 72–14420. Percorso di Continuita' Assistenziale per anziani ultra 65enni non autosufficienti o persone i cui bisogni sanitari e assistenziali siano assimilabili ad anziano non autosufficiente. Piemonte, Italia Bollettino Ufficiale 2004.

20. Mistiaen P: Hospital discharge: problems and interventions. PhD thesis. NIVEL; 2007.

21. Huang TT, Liang SH: A randomized clinical trial of the effectiveness of a discharge planning intervention in hospitalized elders with hip fracture due to falling. J Clin Nurs 2005, 14:1193–201.

22. Phillips CO, Wright SM, Kern DE, Singa RM, Shepperd S, Rubin HR: Comprehensive discharge planning with postdischarge support for older

patients with congestive heart failure: a meta-analysis. JAMA 2004, 291:1358–1367.

23. Parker SG, Peet SM, McPherson A, Cannaby AM, Abrams K: A systematic review of discharge arrangements for older people. Health Technol Assess 2002, 6:1–183.

24. Shepperd S, Parkes J, McClaren J, Phillips C: Discharge planning from hospital to home. Cochrane Database Syst Rev 2004, 1:CD000313.

25. Richards S, Coast J: Interventions to improve access to health and social care after discharge from hospital: a systematic review. J Health Serv Res Policy 2003, 8:171–9.

26. Godfrey M, Keen J, Townsend J, Moore J, Ware P, Hardy B, West R, Weatherly H, Henderson C: An Evaluation of Intermediate Care for Older People. In Final Report. University of Leeds, Leeds; 2005.

27. Plochg T, Delnoij DM, Kruk TF, Janmaat TA, Klazinga NS: Intermediate care: for better or worse? Process evaluation of an intermediate care model between a university hospital and a residential home. BMC Health Serv Res 2005, 5:38.

28. Trappes-Lomax T, Ellis A, Fox M, Taylor R, Power M, Stead J, Bainbridge I: Buying Time I: a prospective, controlled trial of a joint health/social care residential rehabilitation unit for older people on discharge from hospital. Health Soc Care Community 2006, 14:49–62.

29. Jackson MF: Use of community support services by elderly patients discharged from general medical and geriatric wards. Journal of Advanced Nursing 1990, 15:167–175.

30. Williams EI, Fitton F: Use of nursing and social services by elderly patients discharged from hospital. British Journal of General Practice 1991, 41:72–75.

31. Mamon J, Steinwachs DM, Fahey M, Bone LR, Oktay J, Klein L: Impact of hospital discharge planning on meeting patient needs after returning home. Health Services Research 1992, 27:155–175.

32. Rorden JW, Taft E: Discharge planning guide for nurses. Philadelphia: W.B. Suanders Company; 1990.

33. Fleming SA, Blake H, Gladman JR, Hart E, Lymbery M, Dewey ME, McCloughry H, Walker M, Miller P: A randomised controlled trial of a care home rehabilitation service to reduce long-term institutionalisation for elderly people. Age Ageing 2004, 33:384–390.

34. Haggerty JL, Reid RJ, Freeman GK, Starfield BH, Adair CE, McKendry R: Continuity of care: a multidisciplinary review. BMJ 2003, 327:1219–1221.

A Systematic Review and Meta-Analysis of the Effects of Clinical Pathways on Length of Stay, Hospital Costs and Patient Outcomes

Thomas Rotter, Joachim Kugler, Rainer Koch, Holger Gothe, Sabine Twork, Jeroen M. van Oostrum and Ewout W. Steyerberg

ABSTRACT

Background

To perform a systematic review about the effect of using clinical pathways on length of stay (LOS), hospital costs and patient outcomes. To provide a framework for local healthcare organisations considering the effectiveness of clinical pathways as a patient management strategy.

Methods

As participants, we considered hospitalized children and adults of every age and indication whose treatment involved the management strategy "clinical pathways." We include only randomised controlled trials (RCT) and controlled clinical trials (CCT), not restricted by language or country of publication. Single measures of continuous and dichotomous study outcomes were extracted from each study. Separate analyses were done in order to compare effects of clinical pathways on length of stay (LOS), hospital costs and patient outcomes. A random effects meta-analysis was performed with untransformed and log transformed outcomes.

Results

In total 17 trials met inclusion criteria, representing 4,070 patients. The quality of the included studies was moderate and studies reporting economic data can be described by a very limited scope of evaluation. In general, the majority of studies reporting economic data (LOS and hospital costs) showed a positive impact. Out of 16 reporting effects on LOS, 12 found significant shortening. Furthermore, in a subgroup-analysis, clinical pathways for invasive procedures showed a stronger LOS reduction (weighted mean difference (WMD)—2.5 days versus—0.8 days).

There was no evidence of differences in readmission to hospitals or in-hospital complications. The overall Odds Ratio (OR) for re-admission was 1.1 (95% CI: 0.57 to 2.08) and for in-hospital complications, the overall OR was 0.7 (95% CI: 0.49 to 1.0). Six studies examined costs, and four showed significantly lower costs for the pathway group. However, heterogeneity between studies reporting on LOS and cost effects was substantial.

Conclusion

As a result of the relatively small number of studies meeting inclusion criteria, this evidence base is not conclusive enough to provide a replicable framework for all pathway strategies. Considering the clinical areas for implementation, clinical pathways seem to be effective especially for invasive care. When implementing clinical pathways, the decision makers need to consider the benefits and costs under different circumstances (e.g. market forces).

Background

Clinical pathways represent a form of "cookbook medicine" that many perceive as an appropriate tool that contributes to quality management, cost-cutting and patient satisfaction.

For the aim of this review, clinical pathways are defined as complex interventions consisting of a number of components based on the best available evidence and guidelines for specific conditions [1]. A clinical pathway defines the sequencing and timing of health interventions and should be developed through the collaborative effort of physicians, nurses, pharmacists, and other associated health professionals [2]. Clinical pathways aim to minimize delays and maximize resource utilization and quality of care [1]. They are also referred to as "integrated care pathways," "critical pathways," "care plans," "care paths," "care maps" and "care protocols."

The effectiveness of clinical pathways is under debate. However, especially in the US, up to 80 percent of hospitals already use clinical pathways for at least some indications [3]. A number of primary studies considered the effectiveness of clinical pathways, but results are inconsistent and suffer from various biases [4-7]. Only one systematic review has been performed, specifically for stroke patients [8]. Narrative reviews are more common, which often rely on "expert opinions" [9-11].

We perform a systematic review and a random effects meta-analysis to assess whether clinical pathways improved the outcome measures "length of stay (LOS)," "hospital costs" and "quality of care" when compared to standard care. By performing a systematic review and meta-analysis we are able to present the available evidence in a substantiated and concise way, in order to provide a framework for local healthcare organisations considering the effectiveness of clinical pathways.

Methods

We followed the methods of the Cochrane Collaboration [12] with some modifications, mainly concerning presentation of meta-analytic results.

Study Selection Criteria

As potential patient samples we considered hospitalized children and adults of every age and indication, whose treatment involved the management strategy "clinical pathways." Given the problem that there are variations in the terminology used in the current research [13], we defined minimum "inclusion criteria" for meeting our clinical pathway definition (see Table 1). Based on our definition (see background), we developed a pre-specified, three operational pathway criteria as follows: 1) multidisciplinary (two or multiple clinical professions involved), 2) protocol or algorithm based (i.e. structured care plan/treatment-protocol or algorithm) and finally, 3) evidence based (pathway components were minimally based on one RCT or best practice guidelines). Every pathway characteristic could be met as (1) "yes" criterion; (2) "not sure" because of poor reporting and the failure to contact the principal author or (3) "criterion not met." If one or more pathway criteria selected is not met, then we excluded the study.

Table 1. Pathway characteristics and quality outcome measures of studies included

Pathway	Characteristics			Quality Measure	Pathway [n/N]	Control [n/N]
Study-ID	multi-disciplinary	evidence-based	protocol/algorithm based		Counts and rates are presented in natural units and as percentages as far as reported	N = numer of participants n = number of events (%) = percentage
Invasive Care						
Grimes, CL	X	X	X	In-hospital complications	20/237 (8.4%)	20/234 (8.5%) N.S.
1998				Re-hospitalisation (6 months)	10/237 (4.2%)	9/234 (3.9%) N.S.
Swanson, CE 1998	X	Not sure	X	Hospital mortality	2/38 (5.2%)	2/33 (6.1%) N.S.
				Mean Modified Barthel Index	92.8	85.6 (p < 0.05)
Dowsey, MM	X	X	X	In-hospital complications	10/92 (10.8%)	20/71 (28.1%) (p < 0.05)
1999				Re-hospitalisation (3 months)	1/92 (1.1%)	0/71 (0%) N.S.
Choong, PF	X	X	X	In-hospital complications	10/55 (18.2%)	14/56 (25.0%) N.S.
2000				Re-hospitalisation (28 days)	2/55 (3.6%)	6/56 (10.7%) N.S.
Aizawa, T	X	X	X	In-hospital complications	1/32 (3.1%)	2/37 (5.4%) N.S.
2002				Re-hospitalisation (6 months)	1/32 (3.1%)	0/37 (0%) N.S.
Kiyama, T	X	X	X	In-hospital complications	3/47 (6.4%)	5/38 (13.2%) N.S.
2003						
Hirao, M	X	X	X	In-hospital complications	19/53 (35.8%)	17/50 (34.0%) N.S.
2005				Re-hospitalisation (6 months)	0/53 (0%)	0/50 (0%)
Non-Invasive Care						
Falconer, JA	X	Not	X	Mortality (12 months)	N.S.	N.S.
1993		Sure		Re-hospitalisation (12 months)	N.S.	N.S.
				Cognitive and functional scores (0–100)	N.S.	N.S.
				Patient satisfaction	7.7 (SD 2.6)	8.8 (SD 1.7) (p < 0.05)
Gomez, MA	X	X	X	Complete and graded exercise test	44/50 (88.0%)	15/50 (30.0%)
1996						
Roberts, RR	X	X	X	Hospitalised patients as %	(45.1%)	(100%)
1997				Re-hospitalisation (8 weeks)	5/82 (6.1%)	4/83 (4.8%) N.S.
Johnson, KB 2000	X	X	X	Unscheduled clinic visits; no hospital re-admission (2 weeks)	0/55 (0%)	2/55 (3.6%)
Kollef, HM	X	X	X	In-hospital complications	9/239 (3.8%)	13/250 (5.2%) N.S.
2000				Hospital mortality	5/239 (2.1%)	8/250 (3.2%) N.S.

Table 1. *(Continued)*

Morris, TJ 2000	X	X	X	Absolute difference in rates (ARR) between pathway and control		1-sided 95% CI upper limit:
				In-hospital complications	(0.6%)	(4.6%) N.S.
				Re-hospitalisation (6 weeks)	(0.7%)	(3.6%) N.S.
				Mortality (6 weeks)	(-0.1%)	(2.5%) N.S.
Sulch, D	X	X	X	Median Barthel Index Score (26 weeks)	17	17 N.S.
1999				Mortality (26 weeks)	10/76 (13.2%)	6/76 (7.9%) N.S.
Kim, MH	X	X	X	Complications until follow-up (27 days)	1/9 (11.1%)	1/9 (11.1%) N.S.
2002				Re-hospitalisation (27 days)	2/9 (22.2%)	0/9 (0%) N.S.
Chen, SH 2004	X	X	X	Emergency room usage (not comparable with in-hospital complications)	3/20 (15.0%)	13/22 (59.1%) (p < 0.05)
				Re-hospitalisation (3 months)	N.S.	N.S.
Ueui, K 2004	X	X	X	Not reported		

Legend: Every pathway characteristic could be met as (1) "yes" criterion; (2) "not sure" because of poor reporting and the failure to contact the principal author or (3) "criterion not met." If one or more pathway criteria selected before not met, then we excluded the study. Due to poor reporting some quality measures are presented only as percentages or mean scores with or without associated standard deviations (SD). Some quality measures were only reported as statistical significance (i.e. p < 0.05) or not significant (N.S.) and any other data were missing.

Please note, additional information relating to the included studies that matched these requirements or differ from each other, are given in the results section of this review.

The setting definition covered the whole range of services offered by the clinical (out- and in-patient) as well as in the in-patient rehabilitation sector. We only gathered robust evidence and limited our study selection to randomised controlled trials (RCT) and controlled clinical trials (CCT) including methodological quality criteria (please see "quality assessment and data analyses").

We considered every objective economic and patient outcome for inclusion. We pre-defined (1) in-hospital complications as a secondary disease or adverse medical occurrence during hospitalization [14] and (2) we defined re-hospitalization as a readmission within a specified follow up period of an index admission.

Data Sources and Search Strategy

We performed specialised searches of the Medline database (1966–2006), Embase (1980–2006), Cinahl (1982–2006), Global Health (1973–2006), and the specialised Cochrane register (including NHS EED and HTA Database; last update: 13.11.06), not restricted by language or country. We used free text words (tw), medical subject headings (MeSH terms -/-) or exploded MeSH terms for our MEDLINE literature search. This controlled vocabulary was adapted (as much as possible) to the indexation (thesaurus) of all other databases included in this review. We demonstrate our "clinical pathway search strategy" with the MEDLINE inquiry (Table 2).

Furthermore, we employed citation tracking, which examines included studies and previous reviews and contacted investigators to identify any study missed by the electronic searches.

Table 2. Clinical pathway search strategy Ovid Medline: 1966 to November Week 2 2006

1.	Critical Pathways/
2.	(clinical path$ or critical path$ or care path$ or care map$).tw.
3.	exp Guidelines/
4.	Health Planning Guidelines/
5.	Guideline Adherence/
6.	(guideline? adj2 introduc$ or issu$ or impact or effect? or disseminat$ or distribut$)).tw.
7.	nursing protocol$.tw.
8.	(professional standard$ or professional protocol or professional care map).tw.
9.	(practice guidelin$ or practice protocol$ or clinical practice guideline$).tw.
10.	guideline.pt.
11.	or/1–10
12.	exp Hospitalization/
13.	(in-patient or hospitalized or hospitalised or hospitalisation or hospitalization).tw.
14.	exp Outpatient Clinics, Hospital/
15.	in-hospital.tw.
16.	exp Hospital Units/
17.	(Patient Admission or patient readmission or patient readmission or discharge).tw.
18.	or/12–17
19.	11 and 18
20.	randomized controlled trial.pt.
21.	controlled clinical trial.pt.
22.	intervention studies/
23.	experiment$.tw.
24.	pre test or pretest or (posttest or post test)).tw.
25.	random allocation/
26.	or/20–25
27.	18 and 26

Quality Assessment and Data Analysis

For quality of studies, we adhered to the Effective Organisation of Care Group (EPOC) module[15] and defined three risk classes: Class I (low risk of bias), Class II (moderate risk of bias) and Class III (high risk of bias). Two reviewers independently assessed and abstracted data, on the intervention criteria, study characteristics and methodological quality. Any disagreement was discussed with a third reviewer. Studies with a high risk of bias were excluded from the review after documentation. If a primary study did not provide information about the standard deviation, we used the approximative or direct algebraic connection between the stated confidence intervals, or p-values, and the standard deviation and calculated the inverse transformation to the individual or pooled standard deviation [12]. Prior to the actual statistical pooling of the singular effects, an adjustment of the data regarding costs due to inflation and price adjustment (OECD Health Care Price Index) was carried out [16]. We chose the US Dollar (USD) as the basic currency. The year 2000 was chosen as a representative year for inflation and price adjustments or exchange rates.

We used Review Manager (RevMan) of the Cochrane Collaboration to calculate a pooled effect estimate, called weighted mean difference (WMD) [17]. We used a random effects model since this model estimates the effect with consideration to the variance between studies, rather than ignoring heterogeneity by employing a fixed effect model. The effect sizes were generated using a model fitting inverse variance weights [17].

Heterogeneity and Meta-Analysis

Despite the expected clinical heterogeneity (clinical variability of the included pathway interventions) within the review, it is important to assess the comparability of the results from individual studies. A useful statistic for quantifying inconsistency is $I^2 = [(Q \, df)/Q] \times 100\%$, where Q is the chi-squared statistic and df is its degrees of freedom [12]. This quantifies the total variance explained by the heterogeneity as a percentage. We considered an overall test-value greater than 60% to serve as evidence of substantial heterogeneity of a magnitude were statistical pooling is not appropriate.

Sensitivity Analysis

In a sensitivity analysis, both fixed effects and random effects models were employed to determine the causes of heterogeneity and test the confidence that can be placed in both estimates. Only robust estimations of the pooled effects with similar results in fixed effects and random effects models are included in the meta-analysis and discussed. Furthermore, sensitivity analyses were performed to test whether the effect size varied by the countries where the study was carried out (adjusting for market forces) and the year of publication, adjusting for temporal trends.

Subgroup Analysis

We decided previously to perform a subgroup analysis of invasive versus non-invasive clinical pathways, whereupon the distinction between invasive or non-invasive interventions refers to the nature of patient management guided by clinical pathways (i.e. clinical pathways for gastrectomy; transurethral resection of the prostate; hip and knee arthroplasty; percutaneous transluminal coronary angioplasty; etc. versus clinical pathways for asthma care; stroke; pneumonia; etc.). According to theories on health economics, invasive procedures can be standardized more easily than treatment strategies in conservative sectors due to the lower treatment variance [18].

Assessing Publication Bias

We used a funnel plot analysis to assess publication bias, i.e. the bias caused by a lower likelihood of publication for non-significant studies. The funnel plot is a scatter-plot with the x-axis representing the effects estimated from the primary studies, and the y-axis representing a measure of the sample size in each study (SE; standard error of the mean) [19]. If publication bias is absent, the diagram shows an inverted symmetrical funnel.

However, given that publication bias poses a threat to the validity of this meta-analysis and the graphical method is subjective in nature, we also applied a statistical approach, often called fail-safe N test.

This test provides an estimate of the number of unpublished or in this context called file-draw studies (Nfs) having an average of no effect that is necessary in the analysis for reducing the pooled effect size from significant to non-significant [20]. Using a critical d level (d crit) of -0.20, the estimate of unpublished studies was calculated with

$$Nfs = (Ntotal (mean\ d - d\ crit))/d\ crit;$$

where Ntotal is the total number of studies included in the Meta-Analysis and d is the overall pooled effect (WMD).

Secondary Analyses

The distribution of the length of stay is limited downwards in a natural way (because of the minimum value is always 0), whereas upwards the values can scatter significantly. According to logarithmic transformation, values with this characteristic can have an approximately normal distribution [21Data analyses were performed using SPSS version 15.0 [22].

Results

Search Strategy and Intervention Characteristics

The specialised search strategy led to the initial selection of 2,386 studies, whereas only 17 matched our methodical requirements (see Figure 1 & 2). For the first stage of the study assessment, we scanned all of the 2,386 titles and abstracts for inclusion, the remaining 256 possibly relevant studies were retrieved as full text articles. Based on the full text assessment, we excluded 190 studies out of 256 because they failed to meet our pathway definition. The majority of the excluded studies failed to meet the multidisciplinary pathway criterion, i.e. it was a therapy guideline issued by a medical association or it was a uni-professional nursing care plan. Others did not meet the "algorithm or protocol based" criterion because there was no structure and detailed care plan. For example a poster with issued guidelines was posted in the Emergency Department.

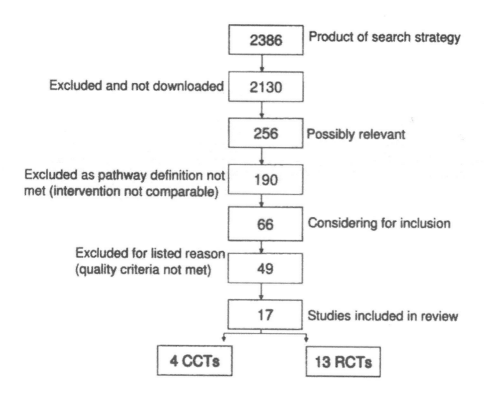

Figure 1. Identification of relevant studies/trail flow.

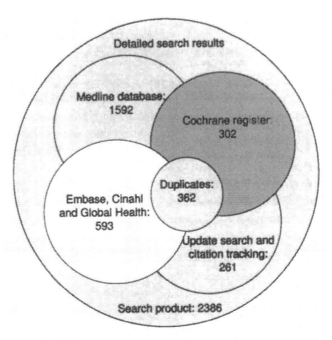

Figure 2. Electronic search results.

Evidence for meeting the minimal criterion "pathway content is minimally based on one RCT" was reported in 15 studies out of 17, which were included in the review. The study from Falconer et al. and Swanson at al. met the evidence criterion, "not sure" because we failed in contacting the principal investigators [23,24].

Intervention Characteristics

The reported pathway strategies can be described as complex pathway interventions versus a "non-intervention" control group or often poorly, described as "usual or traditional care" group. Most of the experimental interventions were combined with other types of interventions like audit and feedback, educational meetings, and reminders. For 8 out of the 17 (47%) included studies, it was clear that the structured care plan was combined with a "clinical diagnostic or assessment protocol"[23,25-31].

The evidence base for two (12%) pathway interventions was "not sure," whereas the remaining 15 interventions were minimally based on one randomized study or good evidence. The reported purpose of the pathway strategies was appropriate management or cost containment.

The hospital setting was in two studies a multi-center study comprising a range of hospitals included in the investigation [26,30]. From the 15 remaining single center studies, 8 studies (53%) were carried out in a university (teaching) hospital setting [24,27-29,32-35] and 7 studies (46%) in a non-university hospital setting [23,25,31,36-39].

Details of the intervention characteristics are given in Table 3.

Table 3. Characteristics of studies included

Study-ID	Study Quality	Country	Sample Size [N]	Mean Age [Years]	Diagnosis/Intervention
Invasive Care					
Grines, CL 1998	Class I	USA*	471	56	Primary Angioplasty in Myocardial Infarction
Swanson, CE 1998	Class II	Australia	67	55	Femoral Fractures
Dowsey, MM 1999	Class II	Australia	163	66	Hip and Knee Arthroplasty
Choong, PF 2000	Class II	Australia	111	81	Fractured Neck of Femur
Aizawa, T 2002	Class II	Japan	69	71	Transurethral Resection of the Prostate
Kiyama, T 2003	Class II	Japan	85	63	Gastrectomy
Hirao, M 2005	Class II	Japan	103	61	Gastrectomy
Non-Invasive Care					
Falconer, JA 1993	Class II	USA*	121	48	Stroke Rehabilitation
Gomez, MA 1996	Class I	USA*	100	52	Myocardial Ischemia
Roberts, RR 1997	Class II	USA*	165	48	Chest Pain
Johnson, KB 2000	Class II	USA*	110	7	Paediatric Asthma
Kollef, HM 2000	Class II	USA*	489	60	Respiratory Care
Marrie, TJ 2000	Class I	USA*	19***	64	Community-Acquired Pneumonia
Sulch, D 2000	Class II	UK**	152	75	Stroke Rehabilitation
Kim, MH 2002	Class II	USA*	18	48	Atrial Fibrillation
Chen, SH 2004	Class II	Taiwan	42	8	Paediatric Asthma
Usui, K 2004	Class II	Japan	61	48	Community-Acquired Pneumonia

Note: USA* = United States of America; UK** = United Kingdom; 19*** = (19) hospitals at random (1743 Patients)

Quality Assessment

To summarize, we examined the design and study quality of 66 studies, excluding 49 out of the 66 studies because of the high risk of bias (see trail flow, Figure 1).

The patient was randomized to the experimental or control group in 12 out of 13 (92%) RCTs. The randomization process was clear in all such studies and justified by the authors. Referring to the individually randomized and single-center studies, the assessment of protection against contamination of the control professionals remained unclear due to poor reporting. None of the investigators reported protection against contamination (communication between experimental and control professionals) and it is possible that control subjects received the intervention. Only the investigation from Marie et al. used a robust cluster randomized design, with 19 hospitals as unit of allocation [30]. To avoid "unit of analysis error," we conducted the meta-analysis at the same level as the allocation (19 cluster-hospitals = 19 patients).

Poor reporting also lead to difficulties in determining the assessment of the power calculations. For instance, sample-size calculation was unclear for over 60% of the included studies; hence the study sample may not have been sufficiently large. Another problem, due to poor reporting was the selection of comparators. The choice of the comparator (i.e. the control and intervention units were located either in the main building or the east building of the participating hospital) was stated and justified by the authors of the 17 primary studies. However, a clear description of what was meant by traditional care or usual care (control group) would have helped in assessing the relevance of the study to other settings.

Primary studies reporting economic data, can be described by a very limited scope of evaluation, focusing on direct hospital LOS and costs effects, rather than on a full economic evaluation [16]. In Table 3 the quality assessment and characteristics of the 17 studies included in the review and meta-analysis are shown in detail.

Effects on LOS

Out of the 16 studies (12 randomized and four non-randomized studies representing a study population of 4,028 patients) examining the effect of clinical pathways on the length of stay, 12 showed significant effects [23,24,26-39]. However, heterogeneity between studies reporting on LOS was substantial (I2 = 80%) and may refer to both the statistical inconsistency as well as to the varying clinical pathway interventions that were included. As a result, the estimation of an overall pooled effect is not appropriate and in Figure 3, the differences from the individual studies in LOS are depicted together with the corresponding confidence intervals without totals. The reported LOS in Kiyama 2003 was calculated from the day of surgery to the day of discharge [39]. All other studies included in this analysis considered the total LOS.

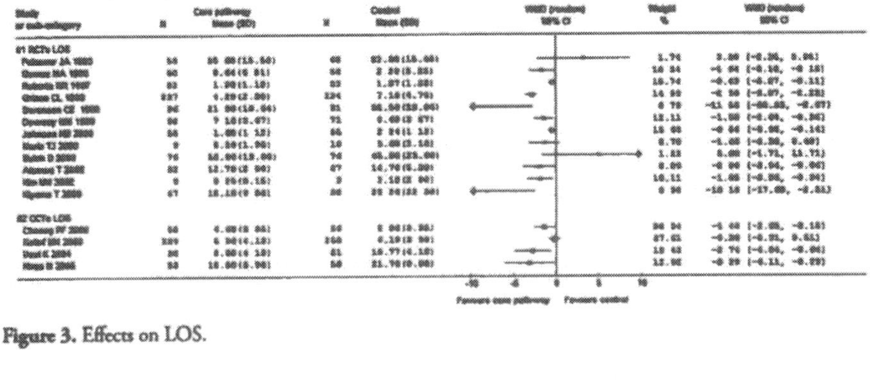

Figure 3. Effects on LOS.

Effects on Patient Outcomes

Out of 17 trials reporting effects on quality outcome measures (see Table 1); six measures were comparable in terms of re-hospitalisation and seven in terms of in-hospital complications [23-35,37-39]. In total, nine primary studies were included in the Meta-analysis (representing a study population of 1,674 patients), examining the effect of clinical pathways on quality patient outcomes. The pooled Odds Ratio (OR) for re-admission was 1.1 (95% CI: 0.57 to 2.08) and for in-hospital complications the overall OR was 0.7 (95% CI: 0.49 to 1.0). Statistical heterogeneity was not present among the studies and there was no evidence of difference in readmission to hospitals or in-hospital complications. The effects of clinical pathways on clinical outcomes in the individual studies are depicted together with the pooled OR (see Figure 4 &5). There was clinical variance in the range of follow-up periods that were used by the investigators measuring re-hospitalization (follow up periods ranged from 27 days to 6 month, see Table 1) as well as the investigators used varying definitions of the term in-hospital complications (included in-hospital complications were cardiac events, infections, thrombosis, re-operation, sepsis and empyema). Obviously, this implies that any time element in the patient outcome data is lost through this approach and it was not possible to compute a series of dichotomous outcomes, i.e. at least one event during the first year of follow up.

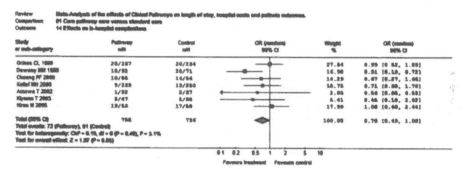

Figure 4. Effects on in-hospital complications.

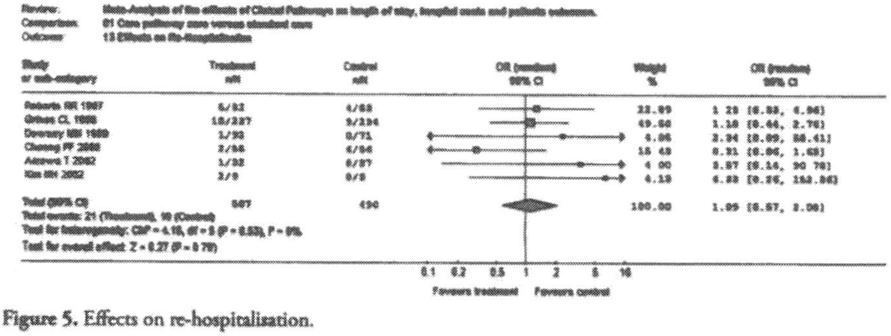

Figure 5. Effects on re-hospitalization.

Effects on Hospital Costs

Six of the included studies (five randomized and one non-randomized), representing a sample of 1328 treated patients, reported on cost effects [26,28,31,33,34,39]. Four out of the five randomized studies found significantly lower hospitalisation costs for pathway groups. The statistical heterogeneity was substantial (I2 = 88%) and compromised the estimation of a pooled effect. Additionally, we also observed a considerable methodological variation which refers to the different methods of cost calculation used by the investigators. Some investigators used a full cost approach (fix and variable costs included), whereas others calculated only direct hospital costs. Table 4 describes the costs differences in detail.

Table 4. Cost data, standardized to the year 2000.

Study ID	Country	Currency	Experiment	SD	Control	SD
Kiyama, T 2003	Japan	US$	$14013	$2634	$18020	$7332
Kim, MH 2002	USA*	US$	$879	$394	$1706	$1512
Kollef, HM 2000	USA*	US$	$922	$1614	$1120	$1430
Grimes, CL 1998	USA*	US$	$11430	$6257	$13733	$7249
Roberts, RR 1997	USA*	US$	$1877	$1243	$2574	$999
Gomez, MA 1996	USA*	US$	$1535	$1985	$6768	$17359

Note: USA* = United States of America

Subgroup Analysis: Invasive Versus Non-Invasive Clinical Pathways

Five of the randomized studies, and two further non-randomized studies assessed the LOS effects of surgical or minimally invasive interventions [24,26,27,29,35,37,39]. The pooled effect for all invasive primary studies was -2.5 days (95% CI: -3.53 to -1.41). The differences in LOS in the individual studies are depicted together with the total effect per study type (RCTs versus CCTs, Figure 6). The statistical pooling of the subgroup of surgical pathway interventions is characterized by a considerable overall test-value for statistical inconsistency (I2 = 60.9%) which also reflects the clinical heterogeneity of the surgical pathway interventions included in this comparison.

The subgroup of the conservative pathway indications [23,28,30-34,36,38] had a reduction of LOS of approximately one day (WMD -0.75; 95% CI: -1.23 to -0.27, Figure 7).

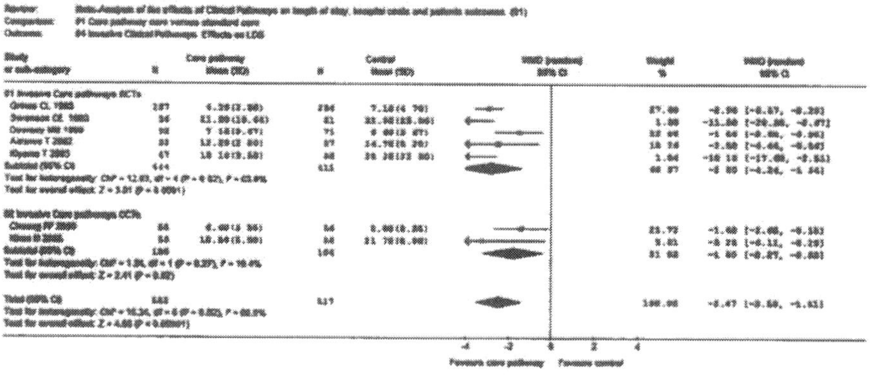

Figure 6. Effects on LOS invasive pathways.

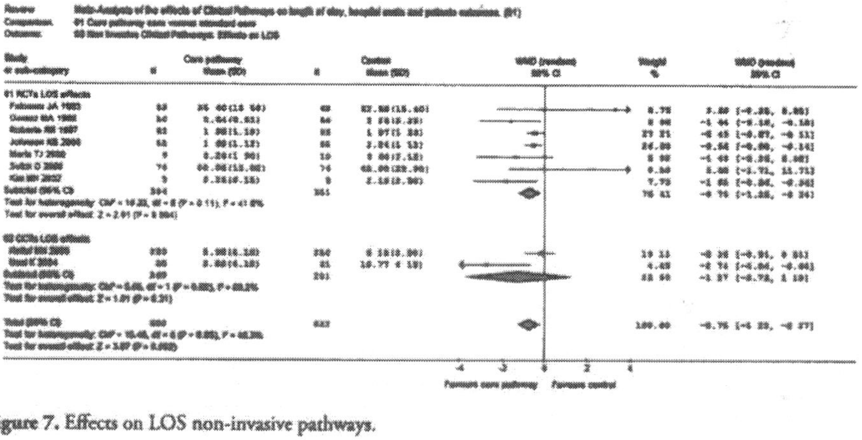

Figure 7. Effects on LOS non-invasive pathways.

Sensitivity Analyses

The LOS effects were robust in terms of the sensitivity analysis concerning the different statistical calculation models (fixed versus random effects model) and the Year of publication, adjusting for temporal trends. However, we observed a trend toward greater reported LOS effects from Japanese studies with a reduction of approximately three days (WMD – 2.7), followed by studies carried out in Australia (WMD – 1.5), Canada (WMD – 1.4) and the USA (WMD – 0.8). Subsequently, we tested the hypotheses, that different market forces (reported effect sizes per country) are possibly confounding the conclusions of these review and meta-analysis. After exclusion (stepwise/iterative and all of the primary Japanese studies) of the subgroup of Japanese studies, the (calculative) overall LOS effect remained robust and statistically significant, but tended to be smaller (WMD – 1.2; subgroup "Japanese studies excluded" versus WMD – 1.5; subgroup "all primary studies" included). This applies also to the subgroup Analysis "Invasive versus non-invasive LOS effects," after exclusion of the subgroup of Japanese studies (WMD -0.6 conservative versus -2.2 invasive).

In addition, the overall odds ratios (OR) for re-admission and in-hospital complications were robust in all terms of the sensitivity analysis, indicating reliable pooled results.

Publication Bias and Other Sources of Systematic Error

The funnel plot showed a relatively symmetric distribution (Figure 8), but the point cloud does not have a distinctive funnel form. The deficient funnel form of the funnel plot can also be due to the relatively high heterogeneity with respect to the different pathway indications of the primary studies included in these review (cross-indicational methodology of the primary studies). Furthermore, the number of studies was relatively small. However, given that publication bias may still exist, the statistical fail-safe N objectively helps to quantify. The calculation about the number of file-drawer studies showed that 101 non-significant studies would have to exist to reduce the (calculative) overall effect size of (WMD) – 1.47 to a mean effect size of – 0.2. These results indicate that unpublished research is unlikely to threaten the validity of the original meta-analysis.

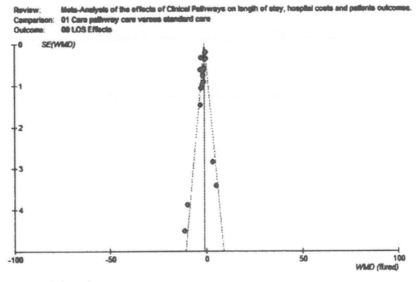

Figure 8. Funnel plot analyses.

Secondary Analyses

The graphic distribution of the original and the logarithmical (natural logarithm LN) LOS data clearly indicates that there is a significant deviation from the normal distribution (Figure 9a &9b). The heterogeneity between studies was not substantially lowered by the log transformation.

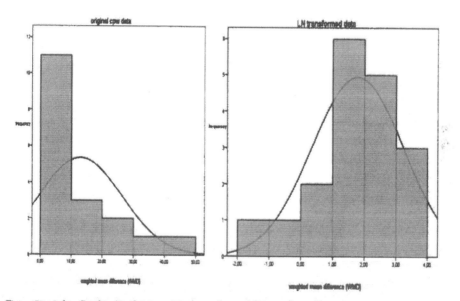

Figure 9. a & b – Graphic distribution original cpw data and LN transformed.

The LOS reduction was estimated as 25 percent (95% CI: -36% to -14%). For example, with the average length of stay of 7 days, the estimated LOS effect was approximately -1.7 days (7 days * -0.25 = -1.75 days LOS reduction). Furthermore, the pooled percentage LOS effects were similar in fixed versus random effects models (WMD -0.25 versus WMD -0.21).

Discussion

In general, the majority of studies reporting economic data (LOS and hospital costs) showed a positive impact. The results suggest that clinical pathways for invasive procedures generate clearer LOS effects (WMD -0.8 conservative versus -2.5 invasive).

Clinical pathways appeared to be effective in reducing LOS and costs. Furthermore, pathways for invasive procedures showed a stronger LOS reduction by comparing the magnitudes of effect. These results may not be applied for acute rehabilitation for stroke, where reverse effects were reported (see effects on LOS, Figure 3) [23,38]. Both trends were not statistically significant but they were in contrast to the majority of pathway effects reported in the present review. However, the question of comparability may rise as a reflection of the differing pathway components included in this review and also applies to the kind and number of providers included in the primary studies.

We did not publish our review protocol prior to the study. The review protocol for the follow-up study will be published as a Cochrane review to prevent any doubt about the comparison to be data-driven instead of protocol-driven. We determined the scope of this review question on a pilot analysis of existing primary study data resulting in a diverse set of included studies. As it is an explanatory analysis, the pooled results of the meta-analyses may only apply for the majority of included pathway conditions reporting positive effects or trends. Another limitation refers to the poorly described control conditions reported in the primary studies and implied both, the risk of contamination, and the masking of effects. Therefore, we did not pool the primary LOS and costs data from all of the 17 included studies and concentrated on examining the relationship between clinical subgroups (i.e. surgical versus non-invasive pathway conditions).

It should be noted that the development and implementation of clinical pathways consumes a considerable amount of resources. This corresponds to the fact that truly achievable costs savings depend on the number of cases (volume). This has to be included in the costs analysis. The inflation-adjusted costs for implementation (without maintenance and further development) of the pathway indication "Caesarian section" amounted to nearly $20,000 [40]. However, since normally 20 percent of the diagnoses cover 80 percent of the cases [18], a considerable percentage of medical services can be dealt with using a relatively small number of clinical pathways. Therefore, the expenditures will amortize rapidly.

It is very important not to look too far into these results, as there were some limitations. Moreover, it has to be emphasised that evidence determined by meta-analysis is always exploratory in nature and should be considered with caution.

Due to the result of the relatively small number of studies meeting inclusion criteria, this evidence base is not conclusive enough to provide a replicable framework for all pathway strategies. Considering the clinical areas for implementation, clinical pathways seem to be effective especially for invasive care. The likely benefits and costs need to be considered by the local healthcare providers when implementing clinical pathways under different circumstances. This review has shown that there is not one, singular strong evidence base. Accordingly, decision-makers should also consider some limitations in relation to the generalization of these findings. Replicating the results of this review in other settings could be problematic (e.g. ceiling effects such as market forces).

The heterogeneity in design and outcomes of the studies was large and refers to the statistical heterogeneity in addition to the clinical variability of the included studies. This precluded the overall pooling of LOS and cost data, although the order of magnitude of effects indicated that there are considerable implications of using clinical pathways.

It is unavoidable that some studies will have been overlooked, despite our electronic search strategy. Studies meeting our clinical pathway definition (see Table 1 & 4) were included, regardless of the fact that the term pathway was mentioned in the study and was done to avoid subjectivity. Also, studies were independently assessed and data extracted by two with any disagreement discussed with a third reviewer.

Finally, should be emphasised that the standard of the primary studies included pose a threat to the validity of the results. While the overall quality of the included studies was moderate, most demonstrated methodological weaknesses such as a small sample size available for analysis.

Conclusion

With respect to the totality of available evidence, the knowledge about the mechanisms through which pathways work is insufficient. Future research should focus on a better understanding of the key elements of clinical pathways that have impact on economic and patient outcomes. It is also surprising that more studies do not consider any cost effects other than those of treatment. Health-economic research should therefore concentrate on costs of development and implementation of clinical pathways.

This investigation is the first systematic review regarding the effects of clinical pathways on process and patient outcomes. We explicitly decided to expand this review and will also include less restrictive study designs in addition to randomized and quasi-randomized trials, to provide a comprehensive theoretical basis. The character of non-experimental studies makes them even more difficult to critically assess and moreover, due to the lack of MeSH terms the search results cannot be as sensitive as those for purely RCT/CCT-based reviews. Another future direction is a more comprehensive, patient-centered approach, concentrating more on patient-outcomes rather then health-economic study endpoints. The next scheduled update for this review is planned for the End of 2009.

Competing Interests

The authors declare that they have no competing interests.

Authors' Contributions

TR: Made substantial contributions to conception and design, acquisition of data, analysis and interpretation of data. In particular, he independently screened all titles and abstracts in the first stage of study assessment and led the study assessment of all full text papers as well as the data extraction. He has also been involved in drafting the manuscript and revised it critically as corresponding author. JK: Made substantial contributions to conception and design, acquisition and interpretation of data. He has also been involved in drafting the manuscript, revised it critically for important intellectual content and given final approval. RK: Made substantial contributions to quality assessment, analysis and interpretation of data. In particular, he also independently screened all abstracts, assessed the full text papers and double checked the extracted study data. Has also been involved in drafting the manuscript, revised it critically for statistical content and given final approval. HG: Made substantial contributions to conception and design, acquisition of data and revised the manuscript critically. He has also been involved in developing the electronic-search strategy and any disagreement between TR and RK was discussed with him as a third party reviewer. ST: Has been involved in acquisition of data, screening of studies, retrieving potentially relevant studies and has been involved in drafting the manuscript. JMO: Has been involved in screening of studies, retrieving potentially relevant studies, drafting the manuscript and revised it critically for statistical content. EWS: Made substantial contributions to acquisition of data, statistical analysis and interpretation of data. Has also been involved in drafting the manuscript, revised it critically for intellectual content and given final approval.

Acknowledgements

This article is based on an oral presentation made at a two day meeting entitled 14th German-Hospital-Controller-Day (14. Deutscher Krankenhaus-Controller-Tag) held in Berlin, Germany in March (29th and 30th) 2007.

References

1. Campbell H, Hotchkiss R, Bradshaw N, Porteous M: Integrated care pathways. BMJ 1998, 316:133–137.

2. Coffey RJ, Richards JS, Remmert CS, LeRoy SS, Schoville RR, Baldwin PJ: An introduction to critical paths. Qual Manag Health Care 2005, 14:46–55.

3. Saint S, Hofer TP, Rose JS, Kaufman SR, McMahon LF Jr: Use of critical pathways to improve efficiency: a cautionary tale. Am J Manag Care 2003, 9:758–765.

4. Porter GA, Pisters PW, Mansyur C, Bisanz A, Reyna K, Stanford P, Lee JE, Evans DB: Cost and utilization impact of a clinical pathway for patients undergoing pancreaticoduodenectomy. Annals of Surgical Oncology 2000, 7(7):484–9.

5. Quaglini S, Cavallini A, Gerzeli S, Micieli G: Economic benefit from clinical practice guideline compliance in stroke patient management. Health Policy 2004, 69:305–315.

6. Roberts HC, Pickering RM, Onslow E, Clancy M, Powell J, Roberts A, Hughes K, Coulson D, Bray J: The effectiveness of implementing a care pathway for femoral neck fracture in older people: a prospective controlled before and after study. Age and Ageing 2004, 33:178–184.

7. Bailey R, Weingarten S, Lewis M, Mohsenifar Z: Impact of clinical pathways and practice guidelines on the management of acute exacerbations of bronchial asthma. Chest 1998, 113:28–33.

8. Kwan J, Sandercock P: In-hospital care pathways for stroke. Cochrane Database Syst Rev 2004, CD002924.

9. Smith TJ, Hillner BE: Ensuring quality cancer care by the use of clinical practice guidelines and critical pathways. J Clin Oncol 2001, 19:2886–2897.

10. Kim S, Losina E, Solomon DH, Wright J, Katz JN: Effectiveness of clinical pathways for total knee and total hip arthroplasty: literature review. J Arthroplasty 2003, 18:69–74.

11. Banasiak NC, Meadows-Oliver M: Inpatient asthma clinical pathways for the pediatric patient: an integrative review of the literature. Pediatr Nurs 2004, 30:447–450.

12. Higgins JPT, Green S: Cochrane Handbook for Systematic Reviews of Interventions 4.2.5. Chichester, UK: John Wiley & Sons, Ltd: The Cochrane Library; 2005.

13. Vanhaecht K, De Witte K, Depreitere R, Sermeus W: Clinical pathway audit tools: a systematic review. J Nurs Manag 2006, 14:529–537.

14. MedlinePlus[internet]: (MD)Bethesda: National Library of Medicine (US). [cited 2005 August 11]. [http://www.nlm.nih.gov/medlineplus/mplusdictionary.html]

15. Bero L, Grilli R, Grimshaw JM, Mowat G, Oxman A, Zwarenstein M: Cochrane Effective Practice and Organisation of Care Review Group (Cochrane Group Module). Edited by: Bero L, Grilli R, Grimshaw JM, Mowat G, Oxman A, Zwarenstein M. Oxford: The Cochrane Library; 2001.

16. Drummond MF, Jefferson TO: Guidelines for authors and peer reviewers of economic submissions to the BMJ. The BMJ Economic Evaluation Working Party. BMJ 1996, 313:275–283.

17. Review-Manager: (RevMan) [computer program] 4.2 for Windows edition. Copenhagen: The Nordic Cochrane Centre, The Cochrane Collaboration; John Wiley & Sons, Ltd; 2003.

18. Schlüchtermann J, Sibbel R, Prill MA, Oberender P: Clinical Pathways als Prozesssteuerungsinstrument im Krankenhaus. In Clinical pathways: Facetten eines neuen Versorgungsmodells. Edited by: Oberender P. Stuttgart: Kohlhammer Verlag; 2005:43–57.

19. Berry DA: Meta-Analyses in Medicine and Health Policy. New York; Basel: Marcel Dekker, Inc; 2000.

20. Soeken KL, Sripusanapan A: Assessing publication bias in meta-analysis. Nurs Res 2003, 52:57–60.

21. Marshall A, Vasilakis C, El-Darzi E: Length of stay-based patient flow models: recent developments and future directions. Health Care Manag Sci 2005, 8:213–220.

22. SPSS: Statistical Product and Service Solutions (SPSS) for Windows. Version 15.0 edition. Chicago, IL 2006.

23. Falconer JA, Roth EJ, Sutin JA, Strasser DC, Chang RW: The critical path method in stroke rehabilitation: lessons from an experiment in cost containment and outcome improvement. QRB Qual Rev Bull 1993, 19(1):8–16.

24. Swanson CE, Day GA, Yelland CE, Broome JR, Massey L, Richardson HR, Dimitri K, Marsh A: The management of elderly patients with femoral fractures. A randomised controlled trial of early intervention versus standard care. Med J Aust 1998, 169:515–518.

25. Chen SH, Yeh KW, Chen SH, Yen DC, Yin TJ, Huang JL: The development and establishment of a care map in children with asthma in Taiwan. J Asthma 2004, 41:855–861.

26. Grines CL, Marsalese DL, Brodie B, Griffin J, Donohue B, Costantini CR, Balestrini C, Stone G, Wharton T, Esente P, Spain M, Moses J, Nobuyoshi M, Ayres M, Jones D, Mason D, Sachs D, Grines LL, O'Neill W: Safety and cost-effectiveness of early discharge after primary angioplasty in low risk patients

with acute myocardial infarction. PAMI-II Investigators. Primary Angioplasty in Myocardial Infarction. J Am Coll Cardiol 1998, 31:967–972.

27. Dowsey MM, Kilgour ML, Santamaria NM, Choong PF: Clinical pathways in hip and knee arthroplasty: a prospective randomised controlled study. Med J Aust 1999, 170:59–62.

28. Gomez MA, Anderson JL, Karagounis LA, Muhlestein JB, Mooers FB: An emergency department-based protocol for rapidly ruling out myocardial ischemia reduces hospital time and expense: results of a randomized study (ROMIO). J Am Coll Cardiol 1996, 28:25–33.

29. Aizawa T, Kin T, Kitsukawa S, Mamiya Y, Akiyama A, Ohno Y, Okubo Y, Miki M, Tachibana M: [Impact of a clinical pathway in cases of transurethral resection of the prostate]. Nippon Hinyokika Gakkai Zasshi 2002, 93:463.

30. Marrie TJ, Lau CY, Wheeler SL, Wong CJ, Vandervoort MK, Feagan BG: A controlled trial of a critical pathway for treatment of community-acquired pneumonia. CAPITAL Study Investigators. Community-Acquired Pneumonia Intervention Trial Assessing Levofloxacin. JAMA 2000, 283:749–755.

31. Roberts RR, Zalenski RJ, Mensah EK, Rydman RJ, Ciavarella G, Gussow L, Das K, Kampe LM, Dickover B, McDermott MF, Hart A, Straus HE, Murphy DG, Rao R: Costs of an emergency department-based accelerated diagnostic protocol vs hospitalization in patients with chest pain: a randomized controlled trial. JAMA 1997, 278:1670–1676.

32. Johnson KB, Blaisdell CJ, Walker A, Eggleston P: Effectiveness of a clinical pathway for inpatient asthma management. Pediatrics 2000, 106:1006–1012.

33. Kim MH, Morady F, Conlon B, Kronick S, Lowell M, Bruckman D, Armstrong WF, Eagle KA: A prospective, randomized, controlled trial of an emergency department-based atrial fibrillation treatment strategy with low-molecular-weight heparin (Structured abstract). Ann Emerg Med 2002, 40:187–192.

34. Kollef MH, Shapiro SD, Clinkscale D, Cracchiolo L, Clayton D, Wilner R, Hossin L: The effect of respiratory therapist-initiated treatment protocols on patient outcomes and resource utilization. Chest 2000, 117:467–475.

35. Choong PF, Langford AK, Dowsey MM, Santamaria NM: Clinical pathway for fractured neck of femur: a prospective, controlled study. Med J Aust 2000, 172:423–426.

36. Usui K, Kage H, Soda M, Noda H, Ishihara T: Electronic clinical pathway for community acquired pneumonia (e-CP CAP) (Structured abstract). Nihon Kokyuki Gakkai Zasshi 2004, 42:620–624.

37. Hirao M, Tsujinaka T, Takeno A, Fujitani K, Kurata M: Patient-controlled dietary schedule improves clinical outcome after gastrectomy for gastric cancer. World J Surg 2005, 29:853–857.

38. Sulch D, Perez I, Melbourn A, Kalra L: Randomized controlled trial of integrated (managed) care pathway for stroke rehabilitation. Stroke 2000, 31:1929–1934.

39. Kiyama T, Tajiri T, Yoshiyuki T, Mitsuhashi K, Ise Y, Mizutani T, Okuda T, Fujita I, Masuda G, Kato S, Matsukura N, Tokunaga A, Hasegawa S: Clinical significance of a standardized clinical pathway in gastrectomy patients (Structured abstract). J Nippon Med Sch 2003, 70:263–269.

40. Comried LA: Cost analysis: initiation of HBMC and first CareMap. Nurs Econ 1996, 14:34–39.

Copyrights

Index